**Guide to Non-Ferrous Metals
and their Markets**

Guide to Non-Ferrous Metals and their Markets

John Edwards and Peter Robbins

Kogan Page, London/Nichols Publishing
Company, New York

Copyright © Kogan Page Limited 1979
First published in Great Britain in 1979 by
Kogan Page Limited, 120 Pentonville Road, London N1 9JN
ISBN 0 85038 159 2

First published in the United States of America in 1979
by Nichols Publishing Company, Post Office Box 96, New York
NY 10024

Library of Congress Cataloging in Publication Data
Robbins, Peter
    Guide to non-ferrous metals & their markets
    1. Nonferrous metal industries. 2. Market surveys.
    I. Edwards, John, joint author. II. Title.
    HD9539. A2R6 1979 380. 1'45'669 78-20656
    ISBN 0-89397-050-6

Printed in Great Britain by
Anchor Press, Tiptree, Essex

Contents

Appendices

List of diagrams and price graphs

Acknowledgements

The authors would like to record their thanks for the help provided in preparing this book notably to Ethel de Keyser, also to David Young, Bronwen Morgan and Leila Kirkham, all at Kogan Page, and to Michael Tout. We also wish to thank the Metal Bulletin for providing statistics.

Part 1:
Production, Development and Sale

Introduction:
The Role of Non-Ferrous Metals

Of the 100 or so known elements in the periodic table, over 80 are metallic. Each metal has its own definite physical characteristics but not all metals have found an important role in industry. This may be because no use has yet been found for them, or that another cheaper metal with similar properties is employed, or that the metal is so poisonous that it cannot be mined, processed, or handled without environmental hazards.

Conversely, there are some metals, even quite rare ones, that have achieved such a massive importance in technology that they are indispensable in modern civilization.

This book is concerned with metals that are commercially important enough to sustain a regularly traded market but it excludes the precious metals, that is gold, silver and the platinum group metals. The markets of these metals are governed as much by currency fluctuations and investment considerations, as by supply from mines and consumption by industrial consumers.

It is hoped that precious metals will be the subject of a future book in this series.

Each metal dealt with in this book has its own market idiosyncrasies. For any one who wishes to follow these markets, this book can only be a primer and a guide. In order to understand why a market behaves as it does on a given day the various metal trade periodicals need to be studied and there should be regular exchanges with those merchants who daily trade the particular metal.

Price influences

Prices are affected by demand, supply, exchange rate fluctuations, strikes, wars, technical innovations, environmental considerations, the price of fuel, changes in stockpile limits, new mines, freight rates, and perhaps more important than all these, in the short-term, the judgement or even the whim of the merchant, producer, consumer or investor.

The smaller the market for a metal, the more susceptible it is to manipulation. Selenium for instance, has a world production of only about 1,000 metric tons. At $12.00 per lb it would theoretically cost less than $30 million to buy the entire world output of this metal with no chance of production being increased because selenium is a by-product of copper production. Selenium is essential in the production of photocopying machines amongst other things. Could it be that for the comparatively small investment of $30 million, an investor could threaten to bring to a halt the world's photocopying machine industry, or at least force that industry to pay almost any price for its selenium supplies?

In practice, this could not quite happen because of the difficulty of buying enough metal without others knowing about it, but by buying 20 or so tons of selenium at a time, the market would go up by at least $1.00 per lb. This type of risky speculation has taken place many times particularly in periods of natural shortage.

Most of the metals dealt with in this book, however, are not by-products. They are mined in their own right but this has not freed the markets for many of these metals from short-term manipulation by merchants and producers alike.

Surprisingly perhaps, there is little investment interest in metals that are not traded formally on one or other of the world's commodity markets. This may be partly due to the volatility of the markets or the specialized knowledge required to deal successfully.

In the copper market for example, non-metal trade investors including banks, insurance companies and pension funds, are prepared to support the market when its value goes too far below cost price. This is because copper is traded on a futures market which has had a good record of utilizing available investment funds.

There is, however, next to no chance of including most of the rare metals in such formal commodity exchanges in the foreseeable future. The worldwide turnover is far too small to justify the cost of such an operation.

The markets of some of the rarer metals are greatly distorted in relation to the true supply/demand factors. These metals are by-products of the productions of other less rare metals, which means, that however great the demand for the by-product, production cannot be increased to meet this demand without a similar rise in demand for the major product.

There is an apparent limit to this state of affairs when the value of the by-product exceeds the value of the major product in a given volume of ore, but this is an extremely rare occurrence as most by-product metals only constitute a small percentage of the major production.

The uses to which these metals are put, are often extremely sophisticated. They may be part of an alloy forming a minute but integral part of a jet engine or chemical process or an electronic device in which the value of the metal used represents an insignificant part of the cost of the entire system in which it is used. This means that if no substitute is available, the price of the metal can rise by several hundred percent before it has any effect on the competitiveness of the consumers' product.

These factors help to create erratic, volatile markets.

In times of high demand, producers are unable to increase output and consumers are not concerned with the price they pay. In the recent past, the price of cobalt, cadmium, bismuth and selenium, to give typical examples of by-product metals, have displayed price rises of around 500 per cent in times of high demand which have lasted for less than six months at a time. These metals have markets that are more volatile than any other commodity or investment market in the world.

To add to the difficulties of predicting future consumption and the uses of such metals, there are further factors to consider. For instance, technicians are inventing new uses for rare metals every day, particularly in the electronics and chemical industries. An obscure piece of research could one day result in an important new use for a rare metal as a semi-conductor or as a catalyst. Furthermore, reserves of metal in the earth's crust are finite and the costs of extracting, refining and distributing metals increase every year, with a consequent increase in the risk of investment in metal production.

Alloys

Very few metals are used in their pure state. They are used instead as alloys, that is in a combination of two or more metals. There is an almost limitless permutation of alloys but the most important of these are those consumed in massive quantities by the steel industry. These are known as bulk ferro-alloys and are the alloys of iron with chrome, silicon and manganese. The markets for each of these alloys are so large that they exceed the markets for most pure metals and represent the most commonly traded form of the three metals concerned. Other ferro-alloys such as ferro-nickel, ferro-titanium and ferro-molybdenum, are also important enough to have their own independent market.

The only other types of alloy to have an international market are those based on zinc, copper, nickel and aluminium.

The future

As with fossil fuels, the demand for non-ferrous metals is growing while their reserves are finite and the cost of production is rising. This obviously means that in the long-term prices must rise.

Between 1975 and 1978 metal prices, with some notable exceptions, have been very low. The prices for copper and nickel have been below cost for many producers. For a number of other metals the price has been so low that investment in exploration and production has proved impossible. This situation is due mainly to the over optimistic expectations of metal producers in the 1960s. The great danger that might result from these circumstances is that when demand does pick up substantially, metal will not be available to meet it, for it can take many years to develop new mines and to build new refineries.

Where the bulk of a metal's production is restricted to only a few countries, or where one country is responsible for over half the world's production, as in the case of cobalt from Zaire or columbium from Brazil, it is likely that the producers will control the price.

Various suggestions have been made to try to alleviate any major disruptions in metal markets. The problem is that such action would have to be taken on an international basis to be effective. Unfortunately the interests of importers and exporters are often diametrically opposed, so that the only action that seems likely to be taken is in relation to the formation of cartels amongst producers and the setting up of stockpile facilities by importing countries. Such arrangements are notoriously difficult to organize and it will probably take a crisis to provide the necessary impetus.

15

This book seeks to explain in a non-technical way how non-ferrous metals are obtained, what they are used for, and how they are sold. It also covers the many influences that affect the supplies and prices of metals. An alphabetical review of the individual metals (pages 61-161), provides a detailed description of the supply and demand position, an historical background of marketing and pricing, as well as a look at probable future trends.

If the supply of metals was suddenly to cease, it would not be possible to sustain life as we know it. This dependence is likely to grow rather than lessen as man discovers more and more specialized ways of using metals and minerals in his quest to advance his standard of living. The manufacture of rival materials, such as plastics, is dependent on metals and the use of plastics to replace metals in certain sectors only enlarges the total market and releases metals for other sectors. Without substitution of this kind there would not be enough materials available at the right price.

It can be argued that it is the extent to which mankind can exploit the mineral and metal resources in the earth's crust that regulates progress in our societies. This simply means that these resources are available at a time when man has the skills and capacity to exploit them. If the consumption of metals on a *per capita* basis was the same in the developing world, notably China, as it is in the richer industrialized western world, the quantities required would increase far beyond the present levels.

Mining, processing, manufacture

The mining, processing and manufacture of metals is second only to agriculture worldwide as a provider of employment and earnings. There is a lengthy process between the extraction of metals from the earth to their eventual use in an industrial or consumer product which involves a host of ancillary industries. Before mining the metal, exploration is required to find suitable deposits and their viability needs to be evaluated.

Mining requires a range of machinery, equipment and provision of an infrastructure, including the possible building of roads, railways and complete townships. Once mined, the next stage is the processing of the ore into concentrates, the smelting and refining of the metals into the form required followed by the shipping to the point of consumption.

Subsequent stages are the marketing and trading of metals involving financial institutions of all kinds; the fabrication of the metals into semi-manufactured form, into alloys or into the many other forms required before it is included in the final product that in turn is distributed and marketed. All these stages involve the provision of employment and earnings.

Mining is no longer under the control of a relatively few powerful groups, based in the richest countries. In the communist bloc, which still has vast untapped resources, it is entirely under state control. In most other countries, the host government either directly controls mining activities or has a strong interest, and is the final authority over the policy and behaviour of the mining companies. This increased state control, which often extends far beyond mining and is carried right up to the sale of the metal to the manufacturing industry, means that a completely new set of criteria is replacing the old economic law of supply and demand. The national and political interests of the country concerned are given priority over commercial considerations. Developing countries, for example, wish to retain more and more of the processing of their raw material resources in the country of origin to provide more employment and earnings; so the first stage is the shift of smelting and refining facilities away from the industrialized countries — already beset by environmental problems — to the developing areas anxious to create industry. The next logical sequence would be for the semi-manufacturing units to follow, and so on. If this trend continues it will cause a considerable disruption in the present pattern of trade.

Changes in the world metal trade

Developing countries, too, in their quest for a new world economic order are seeking to control sales and marketing of metals to a much greater extent than in the past. With increasing wealth, and with populations growing at a faster rate than in the present industrialized countries, it is likely that the main growth in demand for metals will be concentrated in the developing countries. Their governments will be expected to give special priority to their domestic markets, which will mean a lesser amount available for export.

Linked by a common desire to earn more for their raw material resources, that were sold too cheaply in the past, it is likely that fresh attempts will be made to launch producer cartels in order to control metal supplies and prices in a similar manner to that employed by the oil producing countries. Although such cartels are unlikely to succeed in the long-term, they could be a powerful influence in times of shortage, when a lack of investment as a result of fears about the future, starts to make itself felt.

The inflation in costs, aggravated by the need to exploit deposits in more remote areas and with less rich ore bodies in consequence of the exhaustion of existing mines, presents formidable problems. It is now unlikely that any single mining group, however prosperous, will be able to finance the exploitation of a major project single-handed, especially in view of the political risks involved and the increasing demands of host governments.

International consortia, and perhaps the creation of an international fund, will be the new order of the day. The days when new mines could be funded by money loaned against forward sales at a known price are also numbered. Inflation, and fluctuations in the currency rates, make fixed price contracts for forward sales almost impossible, without the inclusion of a host of sub-clauses to deal with all possibilities.

In any event the sanctity of long-term contracts has been brought into disrepute. Governments have refused to honour them either because of a change of government, a change of attitude or simply because changed circumstances mean that the contract conflicts with the national interest of the country concerned.

The need for some form of guaranteed return, and the drive for commodity price stabilization generally as a means of sharing the world's wealth more fairly, are bringing profound changes in the market place. These considerations are

17

overriding the basic law of supply and demand which so far has been the over-whelming influence in the metal markets.

At the same time the growth in international trade has promoted the growth of multinational groups engaged in the buying and selling of metals. Mining companies, faced with governments taking over many of their previous activities are tending to protect themselves by forming larger and more powerful units. They are also expanding vertically into merchanting, broking and metal manu-facturing.

Merchants and broking groups, helped by the vast improvement in communi-cations, have expanded internationally. They are operating in many countries and markets instead of confining themselves to their domestic situation. The expansion of East-West trade has been due largely to merchants seeking new trading opportunities in the communist bloc countries and attention is now being focused on the developing countries, as importers as well as exporters. The aggressiveness of the major Japanese merchanting groups — often acting in concert and backed by government support — has been a prime influence in expanding merchant activity and importance.

Political significance of distribution, production and consumption patterns

Political considerations are now becoming more important in the metals industry in line with the changing forces in the world. At the start of the industrial revolution trading in metals was largely controlled by the consuming countries, who developed and marketed raw material supplies throughout the world.

At one stage Britain, through the Commonwealth and as the largest importer of metals, controlled much of the trade through various big mining companies formed for that purpose. Later rich and powerful mining companies, based in the United States spread their tentacles throughout the world by investing in the exploitation of metals required to fuel the industrialization of the western world. Powerful mining groups based in other countries also operated inter-nationally, although usually, like the US companies, with a large domestic base for their operations.

This method of control is not new. Similar controls on trading in raw materials were exercised by the Phoenecians, the Romans and later the Spanish and Portuguese. They all brought back riches of one kind or another from countries able to supply their needs.

Once again the order is changing. The communist bloc countries give priority to their own needs and ambitions, introducing new concepts into world trade. Other raw material supplying countries have become politically independent and seek to control their own destinies and assets; putting their needs before the requirements of the industrialized world.

Independent developing countries are eager to improve their own standard of living by increased industrialization. They are dependent on exports of metals, and other raw materials, for the bulk of their foreign exchange earnings and more and more metal supplies are likely to come from developing countries as the deposits in the industrialized world become exhausted.

Growing Importance of Developing Countries and the Trend towards International Market Control

There is a growing awareness amongst the countries supplying raw materials that they have been handing over their national resources — unrenewable in the case of metals — to the industrialized countries on terms that have enabled the rich countries to become richer, using resources provided by the poorer countries.

This realization that the industrialized world was waxing fat on raw materials supplied by developing countries, who were becoming poorer as a result, has built up a mood of discontent. It received a further, possibly decisive, push when the oil producing countries' cartel — OPEC, the Organization of Petroleum Exporting Countries — in 1973, exposed the vulnerability of the industrialized world to curbs on a vital raw material and demonstrated that the richer countries could be forced to pay more. Suppliers of other raw materials, including metals, in developing countries were encouraged to emulate the example set by the oil producers and to seek a far greater share of wealth for their resources.

However, it has not proved simple. Oil is in a specially favourable position for a cartel. Its reserves appear to be far more limited than those of many other raw materials, although in the long-term it is just another form of energy, and in the short-term the industrialized world is dependent on oil. There is a common political link between the main oil exporting countries that enables them to cooperate far more easily than is usual for suppliers selling the same product. The geographical and political closeness amongst the oil producers has enabled them to form an effective cartel, at least for the time being, and to resist the economic pressure imposed by the industrialized world.

It is far more difficult for developing countries, dependent on metal exports, to control the markets in this manner. The production of most metals is spread over a wider geographical area; much of the present output of metals is in the industrialized countries where the growth of secondary or scrap metals competes with primary supplies; and large percentages of refining and smelting facilities are also concentrated in industrialized countries, since they are highly capital intensive units requiring large investments.

Of primary importance is the fact that with most metals, and indeed most raw materials other than oil, the main exporting countries find it almost impossible to cooperate effectively. Many of these countries have governments which differ in their political outlook and which tend to regard other exporters as rivals in the market place.

Bearing in mind their newly won independence, and the strong sense of nationalism that helped achieve that independence, it is not difficult to see the

19

problems involved in poorer countries cooperating for the common purpose of controlling a market, in the teeth of powerful opposition from the industrialized world on whom they are still largely economically dependent. Many of the developing countries involved simply cannot afford to cut back production of their main source of export earnings without creating unacceptable unemployment levels, and poverty in their domestic situation. So they remain subject to the economic ups and downs in the industrialized world where fluctuations in demand — as a result of the rise and fall in industrial activity — to a large extent decides the prices paid to the supplying countries for their metals. Thus during an economic recession it is the developing country supplying raw materials that bears the brunt, since it is often unable to adjust output to the lower level required, because of the effect on unemployment and on its foreign exchange earnings.

This vicious circle is completed when surplus production brings prices down to below the cost of output causing heavy losses to the countries least able to afford them. Ironically the success of the oil producers' cartel unleashed the economic recession that demonstrated the fallibility of other raw material suppliers.

Producer cartels

Moves to break out of this stranglehold have been unsuccessful so far. The copper exporters have demonstrated the problems involved in forming a cartel. In 1967 the Intergovernmental Council for Copper Exporting Countries (known by its initials in French, since it is based in Paris, as CIPEC) was formed. It had four founder members — Chile, Peru, Zaire and Zambia — which between them controlled nearly 70 per cent of the world's trade in copper exports, not production, since the biggest producers of copper are the United States and the Soviet Union, who consume their own output domestically.

By 1978 the founder members had been joined by Indonesia as a full member, Australia, Papua New Guinea, Yugoslavia and Mauretania as associate members and Canada as an observer. But during those 11 years, CIPEC totally failed for a lengthy period either to control the copper market or to prevent a slump in prices to well below the cost of production levels, despite accounting for the bulk of the world's copper trade.

Political differences between the member countries have not helped. The government in Chile changed radically during the period, with the replacement of the Allende Marxist government by a military dictatorship to which other CIPEC governments were hostile. There have also been disputes between the other member countries, more interested in pursuing their own interests and faced with differing domestic problems. Two major achievements were the agreement to cut back production by 15 per cent at one stage, and persuading the Japanese Government to stop exporting its surplus copper to the world market and thereby depressing prices.

However, the cutback agreement proved to be more transitory than real, since it was based on expected production in order to allow for expansion plans. Some countries were actually exporting larger quantities after the cutback than before, and the impact on market prices soon faded away and respect for CIPEC also suffered a severe blow. The Japanese maintained their export ban. But this entailed the build-up of a huge stockpile, which simply added to the surplus stocks held around the world that had such a depressing influence on prices.

Having failed to make any progress as a producer cartel, copper exporting countries subsequently looked at an alternative method of stabilizing the market in cooperation with consumers.

The model for this alternative type of producer/consumer cooperation is provided by the International Tin Agreement, which has effectively controlled the tin market for over 20 years. In many ways tin is one of the best candidates for a producer cartel, since output is concentrated in six or seven main producing countries, most of which are developing nations with a common interest in raising the price of tin. In addition several of the most important producing countries — Malaysia, Indonesia and Thailand — are close to each other geographically and face similar political problems. Bolivia, the world's second largest producer of tin, has a constant interest in obtaining price increases since it is a high cost producer compared with Malaysia, having to dig the tin out of mines high up in the unfriendly Andes where oxygen is in short supply and mining conditions, in consequence, very onerous.

There is a long history of control schemes in relation to tin which were originally organized by UK-based mining companies. The price of tin tends to be very volatile and its supply/demand situation is closely linked to trends in the industrialized world through its main outlet, tinplate, which, as part of the steel industry, is highly sensitive to economic changes.

These various schemes met with limited success. They proved costly to operate and did not prevent prices fluctuating wildly. The Tin Producers Association gave way to the International Tin Committee, which attempted to include consumer representatives but they played little part. However, after the Second World War it was decided to make a new effort to control the tin market, combining the forces of the leading producing and consuming counties with equal representation under a new International Tin Agreement. This finally came into force in 1956. Although consumers had equal voting rights, the basic objective of the Tin Agreement is to protect the interests of the tin producing countries. These were clearly set out in the first Agreement, but were enlarged and expanded on in subsequent Agreements, to take account of the many various interests involved.

The basic objectives remain as outlined in the preamble to the first Agreement:

'a) to prevent or alleviate widespread unemployment or under-employment and other serious difficulties which are likely to result from maladjustments between the supply of and demand for tin;

b) to prevent excessive fluctuations in the price of tin and to achieve a reasonable degree of stability of price on a basis which will secure long-term stability between supply and demand;

c) to ensure adequate supplies of tin at reasonable prices at all times.'

Objective (c), subsequently modified and enlarged, was essentially a sop to the consumers since no one has yet devised an effective way of preventing prices from rising once a shortage develops. The basic formula of the Agreement is that a buffer stock is created with the purpose of buying up tin when a surplus situation develops and prices threaten to fall below an agreed 'floor' level at which it becomes uneconomic for producers to continue production. The buffer stock also sells any surplus tin to try and prevent prices going through an agreed 'ceiling' level, above which consumers believe prices are unreasonably high and which will hit demand. Prices for tin were originally the cash quotations on the London Metal Exchange. But in 1972, following the 'float' of sterling, it was decided to use the Straits tin price quoted in Malaysian ringitts as the basis for buffer stock price ranges under the International Tin Agreement. Ideally prices should remain, under the Agreement, in the mid-range at which point the buffer stock does not normally operate. The buffer stock is a net buyer of tin in the lower range, with the floor price at the bottom, in order to prevent the market reaching and going below the floor. The buffer stock manager's task is to decide when and how to start buying, with funds provided by the member countries, should the market enter the lower price range.

Once the floor is reached, the buffer stock is supposed to take in all tin offered to it until its financial resources are exhausted. An added protection against the floor price being breached, as has happened in the past, is the International Tin Council which has the power to impose export quotas on the producing countries when the buffer stock holdings reach a certain level.

The imposition of export quotas is a vital key to protecting the all-important guaranteed floor price for tin which is quoted by producers when seeking funds to expand or maintain production. Quotas provide the fundamental answer to a surplus situation, but they take some time to have an effect and it is therefore important for the buffer stock to retain control in the interim period. For this reason it was agreed in the Fifth International Tin Agreement, which finally came into force in 1977, that the buffer stock should be doubled in size from the equivalent of 20,000 to 40,000 metric tons. The new total is made up of 20,000 metric tons, compulsorily contributed by the producing countries, who gain the main benefit from the protection of the floor price, and 20,000 metric tons of voluntary contributions from the consuming countries. It was argued that a larger buffer stock would enable more holdings to be acquired in times of surplus, leaving a larger quantity to be disposed of in times of shortage, when the ceiling price is threatened.

So far commodity agreements of all kinds, including tin, have failed to find a formula whereby prices can be prevented from going through the ceiling. Money is usually available to protect the floor price, possibly with the help of export or production quotas. But shortages force prices up which usually means that there are inadequate supplies available in the buffer stock to prevent the ceiling being breached. Once the buffer stock holdings are exhausted the market is out of the control of the Agreement. There is a built-in disadvantage to consumers anxious to defend the ceiling in that contributions to the Tin Council buffer stock are normally in the form of money, representing the equivalent tonnage of tin, rather than in the form of actual tin itself. However, in the Fifth

Agreement the United States planned to make its voluntary contribution with surplus tin from its strategic stockpile.

Producing countries, however, prefer to make cash contributions since they can then maintain their trade with consumers in the tin itself at what they hope will be a higher price than the nominal value of their buffer stock contributions. This means that the buffer stock normally starts with no surplus holdings of actual tin, unless these are carried over from a previous Agreement.

The shock administered by OPEC in 1973, and the realization of its growing dependence on suppliers of raw materials, persuaded the United States to join the Fifth International Tin Agreement. The Soviet Union had joined the Fourth Agreement and the only major tin producing or consuming country which is not a member at present is China.

Strategic stockpiles

Prior to the United States joining the Agreement it had maintained a large measure of control over the tin market by building up its strategic stockpile of tin to over 350,000 metric tons. This was equivalent to about two years' world consumption at the time when the stockpiling purchase programme ceased in 1955.

Release of surplus tin from the stockpile in times of shortage helped fill deficits of new production to consumption on several occasions and kept prices under control. The threat of stockpile releases had an equally restraining effect. But on the whole the United States Government cooperated to a large extent with the International Tin Council in, for example, agreeing not to sell stockpile tin below the market price and not to sell it at all when the buffer stock was engaged in support-buying operations.

Nevertheless the threat of stockpile sales of tin by the world's largest consumer kept a tight rein on the demands of producing countries for upward revisions in the Tin Agreement price ranges. Theoretically, the price ranges, apart from being renegotiated with each new Agreement, are also adjusted in line with the trend in production costs, in order to ensure that producers receive a sufficiently reasonable return so that future supplies will be adequate.

In practice, the state of the market was the major element in deciding whether or not the producers pressed for a rise in the price range. Any rise in the price range has to be voted for by both producing and consuming members. This means that the United States, by reason of being by far the largest consumer of tin, has a virtual veto on price range increase proposals, much to the bitterness of producing countries. Since the entry of the United States as a member of the Fifth Agreement, debates for rises in the price ranges have become much more political and contentious.

United States membership of the Tin Agreement marked a change in its general hostility to commodity agreements, which until then it had viewed as a form of producer cartel in direct opposition to its anti-trust laws. It also coincided with the United States taking a fresh look at its stockpile policy. The decision was taken to review the basic objectives of the stockpile. Stockpile releases had already been used to hold back price increases not only in tin but in a variety of other raw materials, including copper and nickel. They were also being used increasingly as a means of raising revenue for the United States Government.

Bearing in mind these depradations of the quantities held in certain commodities, and the concern about the future supply of some raw materials, it was decided to revise the stockpile objectives and to base them on anticipated requirements for a three-year (non-nuclear) war. This review produced a completely new set of objectives, including the building up of certain stockpiles, including copper and nickel, which had been run down. The copper stockpile, which had reached over a million short tons (2,000 lbs each) in 1960, was completely sold out in 1973 when a strike over a wage claim in the United States domestic producing and refining industry lasted nine months and created an acute shortage of supplies.

Another new objective was to cut drastically the stockpile requirements of several metals, including tin and silver. A most important decision was that the stockpile objective for tin was to be cut from 232,000 to 40,000 short tons. The stockpile holdings of tin have been eroded from 350,000 tons in 1955 to around 230,000 short tons in 1968, when stockpile sales were suspended for five years in order to stop conflicting with the operations of the Tin Agreement.

Sales were resumed again in 1973 as part of a new policy of using the stockpile resources to boost national revenue, with Congress authorizing the sale of 30,000 short tons. By 1978 the stockpile authorities had disposed of the 30,000 short tons, which meant that the stockpile had been reduced to just over 200,000 tons, of which some 160,000 tons had been declared surplus to requirements under the revised stockpile policy.

However, attempts to obtain authority from Congress for further releases met with political problems and delays. The Carter Administration on taking over in the United States suspended stockpile activities for a year while it decided whether or not to endorse the new objectives decided under the Ford Administration. It then agreed to the new stockpile objectives but with some reservations.

Many powerful voices in Congress were dubious about the advisability of running the stockpile holdings of tin so low in view of the loss of United States influence in the main tin producing area in the Far East, following the Vietnam War. Allied to this was the fact that tin is one of the few important metals not found in any large quantities on the North American continent. Even the United States voluntary contribution to the Tin Council buffer stock was held up while the debate raged.

Outside the United States, stockpiling of raw materials, especially metals, against the possibility of future shortages, and as a means of stabilizing the markets, has been gaining acceptance. The Japanese and French created their own national stockpiles, since both countries have strong links between the state and metals industries in buying and selling policies.

UNCTAD

On an international scale the United Nations Conference on Trade and Development (UNCTAD) launched a new initiative to try and achieve the new economic order demanded by developing countries. The idea is that by stabilizing commodity prices at higher levels a flow of wealth would be forced from the consumer, industrialized countries to the poorer, supplying countries.

The idea behind the integrated commodities programme, launched by UNCTAD, was for a common buffer stock fund to be created, financed by the richer countries. This is to serve as a back-up for a variety of raw materials where the market needs to be stabilized by artificial means at a higher level in order to provide greater export earnings for the developing countries. Copper is one of the key 10 'core' commodities specified by UNCTAD as in need of an international commodity agreement, on the lines of the tin pact, and backed by the common fund instead of each commodity having its own individual buffer stocks.

At first consumer countries, frightened by the example of OPEC, saw commodity agreements between producing and consuming countries as a useful alternative to producer cartels. But then serious doubts began to creep in about the effectiveness of a common buffer stock and international agreements for certain commodities, including copper. The common fund became the political flashpoint for demands by producing countries for a new economic order giving them a bigger slice of the world's wealth.

Consumer countries argued that a common fund would be extremely expensive, especially if the copper market was included, and probably ineffective in stabilizing markets. They feared that maintaining artificially high prices for a lengthy period would encourage inflation, permit inefficient producers to remain in business and lead to the creation of even bigger, and more costly, surpluses.

Various alternatives have been suggested. These included the creation of national stockpiles, or reserve stocks, held against any shortages of supplies but only loosely linked to any international agreement. Another suggestion was that the export earnings of developing countries should be guaranteed at certain minimum agreed levels instead of the raw materials supplied.

This principle has been incorporated in the Lomé Convention between the European Economic Community and the so-called ACP (African, Caribbean, Pacific) group of developing countries. The idea is that by guaranteeing an agreed export earnings minimum figure − made up by payments from the Community if the country fails to achieve the minimum figure because of low commodity prices or an unpredictable setback in production − the supplying countries will be encouraged to maintain output without the fear of surplus supplies ruining their earnings.

One suggestion is that the EEC as the biggest importer of metals in the world might incorporate copper under the Stabex system of the Lomé Convention. The EEC is also working on plans for an international insurance scheme for mining companies to protect them against developments in supplying countries, which could result in their being nationalized without proper compensation. The threat of this happening, often as a result of a change of government, has brought mining exploration and expansion to a standstill in many areas deemed to be politically unsafe by mining groups and bankers.

Whatever the result of these political confrontations, it is apparent that developing countries which supply the bulk of the world's primary metal exports

25

will play an increasingly important part in controlling the markets. There has already been a wholesale move towards nationalization in most developing countries, either with or without adequate compensation, with the host government playing a far bigger role in the selling and marketing of the production that provides the lifeblood for their economies. This trend will continue as the world becomes more dependent on metals from the developing countries. But the richer industrialized world is unlikely to give in to blackmail by producer countries or cartels. In line with the changing political patterns in the world, non-ferrous metals, as vital ingredients in the process of industrialization and in the acquisition of wealth will be in the forefront of these political and economic struggles.

Future Supplies and Pricing Policies

Much has been made in recent years of gloom and doom predictions forecasting that the world will run out of metals and minerals in the future as known reserves become exhausted. Learned studies have apparently proved conclusively that the growth in consumption of metals is outstripping the discovery of new deposits at such a rate that the world will run out of supplies within the next 100 years (and very much sooner in the case of some metals). One particular example cited was mercury (quicksilver), supplies of which, according to one authoritative prediction should have been exhausted by now.

Certainly the projected growth in demand for metals in the years ahead is frighteningly large, as the world production expands and the demand for better standards of living accelerates.

But the gloom and doom predictions reflect a narrow, naive view. For a start the development of new materials like plastics has already had a profound effect on the consumption of metals and it is reasonable to assume that further developments of this nature will take place in the future. Indeed, optical fibres and micro-electronic 'chips' are existing examples of how industrial societies may be revolutionized and similar developments affecting metals are likely in the near future.

Reclamation

It is often forgotten that many metals, and the products in which they are an element do not disappear once they have been used. In some cases a large percentage of primary metals taken from the ground are used repeatedly as they are virtually indestructible. When metals are turned into chemicals or powders they normally cannot be reclaimed but in most instances metals are recoverable. The supply is simply shifted from below ground to above ground and the 'deposits' above ground are increasing in size all the time.

So far, above the ground 'deposits' have barely been touched. In the case of lead, copper and aluminium, large percentages of total supplies come from secondary or scrap sources in industrialized countries. In these countries, the above-ground deposits have built up over the years and reclaiming techniques have been worked out on an economical basis. In the case of lead, for example, the system created to recover batteries has been highly developed and the bulk of supplies come from secondary rather than primary sources.

In aluminium and copper a large percentage of total supplies come from

27

scrap sources, and there has been considerable growth in the recovery of alloys that can be used again. But recovery of previously used material depends on its being economic and available in sufficient quantities to be competitive — usually cheaper — than the primary product. This means that in many parts of the world where consumption is low there is very little recovery of secondary metals because it is not economically justified.

Nevertheless the supplies are there, and will become recoverable in increasing quantities if the right economic stimulus is provided — in other words if the price paid is right. At the same time increased price incentives would greatly improve the methods of recovering scrap materials, and expand the types of metal reclaimed.

Until now it has not been worthwhile to develop more sophisticated techniques for reclaiming metals. At the moment the reclamation industry relies mainly on crude and simple recovery techniques such as sorting, cutting and the use of a magnet to pick out the steel, as opposed to the non-ferrous parts. Chemical and other modern techniques are not widely employed because they are not often economically viable. But if a shortage of primary supplies did develop, and prices moved to much higher levels, then the economic incentive to recover more scrap metals would obviously become worthwhile and would be exploited. There is a huge hidden supply of reserves in industrialized countries, a fact which is only registered in times of emergency. However, a greater awareness of recovery potential has been building up in recent years. This is linked with the demands of environmentalists for the removal of unsightly waste and for cleaner manufacturing methods.

Pricing

Price is also the key to primary supplies. It should be remembered that in the case of metals, mankind has until now only exploited the one third of the earth's surface that is not covered by the sea. Even that one third has not been fully used; many of the remoter areas like the North and South Pole regions, the Amazon basin and large tracts of Asia and Africa, have not been fully exploited for their mineral and metal reserves. These areas are in many cases remote and difficult especially when there are so many easier deposits available. Over the centuries that mankind has been mining metals, there has been a natural concentration on deposits that are the easiest to exploit since these are the most profitable.

The metal content of the ore is important but by no means the only factor to consider. First the separation of the metals contained in the ore has to be considered. It might be an ore which has too many metals, involving costly separation processes. Ore containing a high proportion of one metal is usually the easiest to handle, since it is then a matter of separating out one metal and throwing the rest away or finding out whether there are some valuable by-products contained in the ore that justify further separation processes.

The miner also has to consider a host of other influences before deciding whether a deposit is worth exploiting: whether there is sufficient water available, how far down the ore is located, the existing infrastructure, available labour, and transport facilities for carrying the ore or the refined metal product to the place where it can be treated or sold. These days an increasingly important factor is the political structure of the host government. The political stability of the government can be crucial in deciding whether or not sufficient funds can be raised to finance the project. In other words economic and political considerations are the prime influences in deciding which metal deposits are to be exploited, not necessarily the richness and size of the ore body. Mining efforts, therefore, have been mainly confined to certain known areas, relying primarily on the improvement in extraction techniques to increase output. Indeed many mines have been worked over several times. The waste product, cast aside by the early miners in favour of richer deposits, becomes financially viable for the more modern miner using a more advanced technique. Thus the percentage of metal required in any ore body to make it worth exploiting progressively reduces and the amount of metal available from any one deposit continues to increase.

When these easy deposits become exhausted, the consequent shortage of supplies can be expected to increase the financial incentive sufficiently to make the more difficult and as yet untapped deposits economically viable, as well as encouraging the miner to go further afield. Additional financial incentives may be required to unlock a large part of the sea bed reserves in view of the different mining techniques that would be required. Nevertheless the supplies are there if needed; some might be available at a far cheaper price once the techniques and political complications surrounding sea bed mining have been clarified. The world debate on dividing up offshore rights is not only concerned with fishing. It is equally concerned with setting the pattern for the future mining of the sea bed. Nodules found on the sea bed, for example, are known to contain large quantities of manganese, copper and nickel as well as many other metals and they are fairly easy to collect.

Looking even further ahead mankind has so far barely scratched the surface of the earth's crust in mining for metals and minerals. Going down deeper into the crust would be far more costly. But if a shortage of supplies was threatened there is no doubt that the ways and means would be found of extracting more of the earth's riches by mining to a deeper level than at present.

Fears of shortages are, therefore, based on possible short-term interruptions of supplies rather than on any fundamental long-term threat. The price at which metals are sold is a complex matter. The basic fundamental, often forgotten, is that the 'real' price of any commodity is what the buyer is prepared to pay for it, not how much it costs to mine or to produce. However, the price that the buyer is prepared to pay can be affected by a host of influences, some logical, others less logical. At the same time it is true to say that the metal cannot go on being sold at below the cost of production for too long without affecting the flow of production and the expansion of output.

There are two basic methods of pricing used for selling metals. One is the so-called producer price system. Under this system the producer fixes the price at which he is prepared to sell, although in fixing the price he has to take into account the state of the market and the attitude of potential buyers, as well as the price of competitive materials and the need to preserve certain outlets.

The producer price system

The producer price system is widely used throughout the world but is most favoured by North American producers and consumers. This is partly because the United States metals industry is more vertically integrated than others. The company mining the metal is very often the same company smelting and refining it. But in the United States the integration is taken several steps further with the mining company owning the semi-fabricator, normally the main buyer, and sometimes also the fabricator's customer, the manufacturer of the final product that is sold to industry or to the general public.

With this kind of captive outlet it is easy to impose a producer price, especially if most of the main competitors in the business have a similar set-up and are therefore likely to have similar cost pressures. Even consumers, who are not owned by mining companies, often favour the producer price system. They believe that it gives them security of supply if the producer is able to charge a reasonable price and as long as the same price is charged to their competitors they need have no fear of being undercut.

At the same time a producer price system normally allows for longer-term price forecasting, since the producer seeks to stabilize the price in his own interests. The producer also smooths out the market place fluctuations by gradual increases to take account of the inflation in his production costs, rather than following the day-to-day changes in the supply/demand position. In other words, the producer is prepared not to capitalize on a sudden shortage situation in the hope that he will not have to sell at a loss if and when there is a surplus supply situation. It is a bargain struck between the producer and consumer, with the producer offering continuity of supply at a foreseeable price linked to the costs of production, and the consumer offering a regular purchase of the metal at a price agreeable to the producer.

The producer price system is used in one form or another for virtually every metal. There are some notable exceptions, such as tin, which is controlled by the International Tin Agreement, tungsten ore and other so-called 'minor' metals where production is scattered amongst many small producers with no leading company powerful enough to set a producer price.

'Free market' system

But the producer price system is often used hand in hand with the other main method of pricing metals — the free market system. Free market prices are essentially defined as those not fixed by producers, although of course as suppliers of the metals, producers must have a considerable influence on the free market value. Free markets tend to be controlled by merchants, who are able to draw on supplies from non-producer price sources. These include communist bloc or other producers not participating in the producer price system; consumers reselling supplies they do not require; secondary or re-refined metals; and purchases made by the merchants from producers in the producer price system possibly at a discount.

Many metals have two prices. The producer price and the free market price.

The free market price is often the price of residual supplies of the metal over and above the regular contracted supplies from the main producer, using a producer price system. In times of shortage, the free market price is above the producer price as buyers seek additional supplies which they are unable to obtain from the producers they normally use. In times of surplus supplies the free market price is at a discount to the producer price in order to attract buying interest.

The importance of the free market in each metal varies considerably. For a start it depends on the share of the total market controlled by the leading suppliers using a producer price system. If, for example, there is a dominant supplier in the communist bloc, who is unwilling to participate in any producer price system adopted by the leading western world producers, then the free market can be most important as a pricing medium for the metal. This is the case with tungsten ore, for which China is the main producer and there is no dominant western world producer able to fix a producer price.

Market situations affecting the normal supply/demand situation can also increase the importance of the free market compared with producer price sales. This was the case with nickel when lengthy strikes at the main producer mines in Canada meant that consumers had to turn to the free market for supplies.

With all metals if there is a shortfall in the normal amount of supplies available from producers under the producer price system, the free market often assumes a more important role. For this reason many consumers, whose factories would be forced to close if they could not obtain supplies of the metals required, like to keep their options open by purchasing a percentage of their supplies from free market sources in addition to their regular supply contracts with producers at the producer price. By so doing they gain some benefit when free market prices are low. They can also expect to be given some sort of priority from their free market suppliers when a shortage situation develops. If this line of communication was not kept open the free market suppliers would be tempted to discard their 'panic' orders when supplies were short or would charge a premium price over that charged to their regular customers.

Nickel provides another example of the changing relationship between the free market and producer price systems. The surge in production of nickel during the mid-1970s, including the entry of several new producers, created a huge surplus of supplies since demand was badly hit by the economic recession triggered off by the 1973 oil crisis. International Nickel of Canada, which at one time accounted for some 80 per cent of total sales and was, therefore, able to fix an effective producer price, was forced in 1977 to abandon quoting official prices in favour of 'confidential' prices in an effort to win back some of the sales lost to its competitors.

Scarcity and surplus

Surplus situations have from time to time created similar distortions in the prices charged for other metals — aluminium and zinc are two such examples.

As regards zinc, the producer price level has had to be adjusted downwards in line with the move in the free market to remain competitive. This was particularly necessary since the producer price is also used as the basis for assessing the

31

cost of ore concentrates bought by the smelters for refining into zinc metal. If heavy discounts are made on the producer price at which the metal is sold, smelters not controlled by mine producers could not afford to buy their raw material intake at a cost based on unrealistically high producer price quotation.

Producers using the producer price system, in fact, can often lose both ways. They are expected, in times of scarcity, to continue supplying consumers at the reasonable contracted prices laid down and not to take too much advantage of scarcity by moving their prices to a much higher level on the free market. However, in times of surplus producers are under great pressure from consumers to reduce their prices in order to be competitive with free market supplies, which buyers fear may enable their rivals to cut manufacturing costs. Faced with these demands the producer is often forced to give sizeable discounts on his official quotation, although he is supposed to stick to the official price during times of shortage.

The prime argument, however, against the producer price system is that it can be inflexible and not anticipate or recognize the changing situation in the market. The price stability provided is in itself a target for competitors to shoot at and also encourages a lack of enterprise by both producers and consumers.

'Free market' trading

In contrast the uncertainty of fluctuating free market prices gives a stimulus to both sides to search for the greatest efficiency. This provides the best protection against competition. The free market, in addition, gives a valuable guide to what is happening, and what is likely to happen, in the market place. In some instances the influence of free market forces is so great that they are dominant over the producer price system. This is the case with copper and lead, as well as with several of the 'minor' metals (that is, metals not traded on the London Metal Exchange) such as tungsten and, nowadays, nickel and quicksilver. Producer prices are fixed in the United States for copper and lead and indeed virtually all metals but this is influenced by movements on the London Metal Exchange markets and on Comex (the New York copper futures market). In the rest of the world the so-called producer prices are in fact based on the London Metal Exchange quotations. The pricing formulae used in the direct supply contracts between copper and lead producers and their customers are based on the Metal Exchange quotations. This is also the case in supply contracts between merchants and their customers, suppliers of copper and lead ore concentrates to smelters and refiners, and to scrap dealers and secondary refiners.

Outsiders find this situation difficult to understand, since it is generally considered that producers — and exporting countries — should be able to fix the prices of their own products and must find it difficult to operate and plan ahead if their returns are subject to wild fluctuations governed by the market place. Similarly consumers find it difficult to plan ahead not knowing the price of raw material intake, which in some cases has a sizeable influence on the cost of the final product, especially for fabricators of semi-manufactured products.

These criticisms of the free market system are all valid. Indeed, there is great concern that the activities of non-trade speculators, whose motives have little connection with the metals industry, should on occasion play a major part in influencing price trends.

There is one basic answer. No one has yet been able to devise a better system for fixing copper and lead prices. There have been several attempts to introduce an international producer price system for copper, first of all by American producers, and later by African and Chilean producers. Possibly the most determined attempt was made in the early 1960s when producers, through market support operations, kept the London Metal Exchange copper price at £234 per ton for nearly two years until the end of 1963. This was followed by a multi-price system whereby the Metal Exchange values were allowed to find their own level, but African and Chilean producers pegged their export prices at a lower level as a safeguard against the substitution of copper by other materials. But this system collapsed in 1966 when Chile decided to increase its export price by £160 to nearly £500 per ton and a few weeks later Rhodesia closed the export route for Zambian copper via the railway to ports in Mozambique. Since that unhappy experience, copper producers have given up trying. The cause has subsequently been taken up by CIPEC (the Intergovernmental Council for Copper Exporting Countries) and the United Nations Conference on Trade and Development (UNCTAD).

But these efforts are aimed more at stabilizing prices, especially with a minimum level, by the use of buffer stock operations to buy up surplus supplies in a similar way to the International Tin Agreement.

The chaos in the market place during the 1970s, as a result of the breakdown of the Bretton Woods fixed exchange rate system and runaway inflation in the industrialized countries, undermined the whole concept of fixed prices, and producers have been forced to make frequent changes to remain competitive with the free market. In 1978 there was a major change when Kennecott, the largest United States' copper producer, decided to abandon their producer price system in favour of basing their prices on the New York copper market (Comex) quotations. A similar change was made by another leading US producer, Anaconda, which meant that other producers were forced to introduce more flexible pricing arrangements for their copper.

Supporters of the free market system for metals argue that it is the fairest way of deciding the 'real' price. They point out that production costs vary enormously from country to country and from company to company, so that unless the producer price is based on the highest cost of production, some producers will inevitably face constant losses. For many minor metals produced as by-products during the processing and refining of the major metal, the price obtained by the producer is based, in effect, on how much he can get. Movements in the price of these metals have no direct impact on the supply situation.

The scrap industry

In the case of copper and lead, as well as aluminium, there is another very important influence to be taken into account — that is the growing secondary and scrap industry. Supplies of scrap metal have a totally different cost structure from primary metals, although the end product may be interchangeable.

Scrap prices are linked closely to primary metals, since they cover virtually the same product obtained from other sources. However, since it is a substitute for the original metal, scrap tends, on the whole, to be cheaper, although this is not always the case. On occasions scrap supplies may be scarce, while primary supplies are abundant and buyers whose processes are geared to using secondary metals might have to pay more. In fact supplies of scrap are also subject to both political and economic influences. Most industrialized countries are reluctant to allow the export of their domestically produced scrap in view of the import savings gained from not having to buy so much primary metal. In consequence, restrictions have been imposed on exports of scrap, which is another reason why it tends to be somewhat cheaper for the domestic consumer who may be the only available buyer. This is also due to the fact that scrap merchants are not equipped with as efficient transport facilities as mine producers who export to a range of countries.

Scrap recovery from the 'mine' above the ground has a totally different cost structure from the mining of primary metals. Scrap refiners rely on the margin between the price at which they buy their raw material intake — either from industrial sources or via collections from the general public — and the price at which they sell the refined product. It is this margin of profit that is all important, not whether the price is high or low. Indeed in boom times scrap refiners often do less well since they have to pay far higher prices for their raw material intake and their profit margin is squeezed compared with more depressed times when the margin can be more easily expanded. During times of economic depression the scrap industry is affected by a cutback in supplies, as opposed to a build-up in primary metals. Reduced industrial activity means less investment by companies who normally generate the bulk of scrap supplies when replacing old equipment with new, or when operating at a higher capacity rate and producing more waste residues.

Aluminium producers, operating on a greater international scale than other metal groups, have managed to acquire and retain a large measure of control over both primary and scrap supplies. This is in part due to the fact that as owners of a large proportion of the fabricating and manufacturing end of the industry, they are also major buyers and, in addition, own a large part of the aluminium secondary refining industry.

But there is not the same integration for copper and lead producers outside the United States and Canada. In the past, lead producers have managed to control the free market prices by support-buying operations, organized amongst themselves. But government disapproval of cartel activities of this kind, especially the anti-trust laws in the United States which are rigorously enforced to prevent collusion amongst producers, has made it more and more difficult for lead producers to operate together, even on international markets.

Copper producers face the same problems of cooperation amongst themselves in a task of much greater magnitude. The quantities and sums involved in the copper industry are very large compared with lead, and the finance required for an effective support-buying policy to control the market would be enormous, and certainly beyond the scope of producers unless backed by government or by international financial resources. Not only is there the cost of actually buying the copper, but also the cost of storage and of having large sums of money tied up for an indefinite period. Many buffer stock operations, including that of the International Tin Agreement, have made money on 'paper'. But a different picture may emerge if the loss from inflation and the lack of interest earned over the period is taken into account.

These hidden costs emphasize the point that influences on the price of metals are not confined to those facing the producer. Political developments worldwide and in specific sensitive areas such as wars, economic trends, interest rates, inflation and changes in currency rates also have to be taken into consideration by buyers and sellers of metals. Reaction to these developments are often more reliably reflected by speculators, who are more sensitive on occasions to the non-trade influences at work. Prices arrived at as a result are the common denominator of all the influences at work, bringing together all the fears and forecasts about future and existing attitudes in deciding the point at which sellers agree to sell and buyers to buy — that is the market price.

The Speculator

There is an understandable concern that excessive speculation may distort the value of metals from the true price demanded by the fundamental supply/ demand situation. This could hurt the producer by forcing prices too low and at other times harm the consumer by raising prices artificially high. This danger has increased enormously in the 1970s with the emergence of huge speculative funds seeking a haven of safety against fluctuations in exchange rates and the ravages of inflation by 'investing' in raw materials, which retain an intrinsic value whatever may happen to the 'paper' money controlled by governments.

Metals are an obvious favourite for these speculative funds, since they do not perish as quickly as other raw materials and can be stored relatively easily and cheaply. Metals also tend to be more closely linked with economic trends than other raw materials, since the demand depends to a large extent on fluctuations in industrial activity.

But the influence of speculators can at times be beneficial to the industry. They provide the 'other' side of the market when the trade might tend to be all sellers, or all buyers, and thus they can be a stabilizing influence. At the same time the inflow of speculative funds into the industry is of great benefit in providing extra financial liquidity, often at a time when the trade is short of the money required to finance its surplus stocks. Otherwise the producer is likely to have to shoulder the main burden since in times of plentiful supply the consumer cuts his stocks down to a minimum, confident that he will easily be able to obtain any extra supplies that might be required.

The speculator also makes possible the efficient operation of the futures market on which the trade can hedge against unpredictable price fluctuations. The fundamental purpose of the futures markets is to provide these hedging facilities. Over the years they have also developed as a price fixing medium for the metals traded.

Futures Markets:
Hedging, Price Fixing and Speculation

Futures markets for base metals are confined to London and New York and to only a few base metals — copper, tin, lead, zinc and aluminium. The collectively defined 'minor' metals, have no central marketing place. Price levels are imposed either by producers or by merchants trading in the free markets amongst themselves, and with both producers and consumers. There is speculation in these minor metals, but it is more difficult than in the 'paper' futures market and is mainly confined to those actually dealing in the metal as a producer or consumer but, most commonly, as a merchant.

Without the benefit of hedging facilities the merchant in particular, has constantly to be alert to changes in the supply/demand position of the various metals. On occasions he has to back his judgement by taking a 'long' or 'short' position that might entail considerable financial risk. He also has to be vigilant to spot occasions when a merchant, a producer or a consumer, attempts to manipulate the market in his favour by buying or selling supplies.

A Minor Metals Traders Association was formed in 1973 by London traders. Several of their European counterparts subsequently joined the Association in an attempt to bring more order to the free markets by establishing some standardization of contracts and of trading regulations. To a large extent trading is dominated by the large groups, often subsidiaries of mining groups and with strong connections on the London Metal Exchange, and with secondary metal dealers.

The mixture of producer prices and merchant-inspired free market trading makes trends in minor metals difficult to follow in what are sometimes highly specialized markets. Fundamentally the trends follow the pattern of industrial and economic activity affecting consumer demand, although movements can often be highly exaggerated.

There are two main futures markets for metals: the London Metal Exchange and Comex. The London Metal Exchange has markets for copper, tin, lead, zinc and silver. It introduced an aluminium contract in October 1978 and was also studying the possibility of a nickel contract. The other main futures market, Comex, is in New York: it has flourishing copper, silver and gold markets, and newly introduced (in 1978), a zinc futures contract.

The two markets strongly influence each other since both in theory and in practice, dealers can make a guaranteed profit by selling in one market and buying in the other if prices move out of line on either side of the Atlantic. These arbitrage operations, as they are known, have to take account of the cost of shipping the metal across the Atlantic, currency exchange rate fluctuations and timing, as well as interest rates and storage facilities. But these are quickly calculated by alert traders and any abnormal difference between London and New York copper or silver prices is usually quickly eliminated by the dealers specializing in arbitrage. Arbitrage operations are frequently 'paper' transactions invoking the possibility of shipping the physical metal rather than actually doing so.

On occasions, such as a dock strike or some other domestic development, prices in the two markets can stay out of line for a longer period if metal supplies cannot be made available to the market where prices are highest. Generally the opening of the Comex market in New York, which corresponds with the early afternoon in Europe, has an important influence on the London market, which in turn affects the trend in Comex. It depends on the stronger influence at the time: whether the New York market is dominant in reacting to some development or rumour emanating from the United States, or whether developments affecting London prices are more powerful. Fundamentally the London Metal Exchange is the more influential of the two markets. Although turnover on the New York copper market is usually larger than in London, it tends to be speculative to a far greater extent. This is because much of the copper traded in the United States was, traditionally, sold under the producer price system. Comex essentially influenced only what is known as 'merchant' copper which is mainly imported together with other supplies of copper acquired by merchants outside the producer price system.

Many of the fabricators, owned by the big mine producing companies, rely almost exclusively on producer price copper. However, they have to watch price movements on Comex to ensure that their competitors both domestically and abroad are not able to buy copper at a sufficiently low price to reduce their manufacturing costs. Traditionally United States domestic buyers of copper have relied on the producers to provide competitively priced copper and, most importantly, a regular flow of supplies under long-term contracts. Comex was regarded as a speculators' gambling den, since a large proportion of the turnover came from the powerful commodity broking firms and commission houses using funds obtained from private and trade speculators who were seeking to make a profit by correctly anticipating the upward or downward move in prices.

However, a major change came in 1978 when the largest United States copper producer — Kennecott — decided to abandon the producer price system and instead base its selling price for copper cathodes on the Comex spot quotation. It charged an additional $2.50 per pound to cover the cost of delivery since the Comex price is ex-warehouse in New York while Kennecott delivers its copper supplies direct to the customer. The company took this radical step to recapture sales lost to cheaper-priced imports. Other producers were, as a result, forced to adapt their pricing policy.

The importance and respectability of Comex has grown in recent years. The United States has gradually become more dependent on imported supplies of copper as a result of consumption growing at a faster rate than domestic production. At the same time the growth of international trade and communications has meant that the United States industry has become more aware of the opportunities outside North America, and the increasing competition facing its manufacturing companies. The activities of speculators have also become more significant with the changes in exchange rates, even of the previously mighty dollar, following the collapse of the Bretton Woods monetary fixed exchange rate system, the oil crisis and the surge in inflation. Political developments outside the United States are also having a much more profound effect on metal prices and supplies. This is reflected in the price movements on Comex.

These influences have, of course, also heightened the importance of the London Metal Exchange as a price fixing medium. The Metal Exchange, which celebrated its centenary in 1977, is the most powerful influence on the prices of copper, lead and to a lesser extent tin and zinc outside North America. Its daily price quotations — usually the settlement price at the close of the morning trading session — are used as the basis for the direct supply contracts between producers and consumers of copper and lead. The quotations are not only used to price transactions in refined metal, but also as a basis for pricing ores, concentrates, semi-manufactured products, manufactured metals and scrap.

The bulk of the international trade in copper and lead is handled through contracts negotiated by the producers with the consumers. But these negotiations usually conducted, in the case of copper, during the 'mating season' — that is in the last quarter of the year — basically deal with the quantity to be supplied during the following year. The price at which the copper is to be bought is based on a formula using the Metal Exchange quotations. This may be, as in Japan and Australia, a producer price that is altered according to changes in the Metal Exchange quotations. Or it may be a much more complicated formula whereby the consumer has a specified period during which he can choose to 'price' the copper being delivered to him using the Metal Exchange quotations during that period. One concession given by producers, for example, known as 'back pricing', is the ability of consumers to fix the price of metals supplied on a chosen date for a limited period prior to the delivery date.

Hedging

An advantage of using the Metal Exchange quotations in these contracts is that the consumer can effectively hedge himself against price fluctuations during the period from which he has priced the copper, to the time when it is manufactured and sold at a previously agreed price.

Producers too can hedge against price fluctuations by making matching purchases of copper committed for future sale, thus ensuring that if the price moves down in the meantime they will benefit from the profit made on the futures market, to compensate for the loss made on the physical metal sold. If the price goes up, then the loss on the futures transaction cancels out the extra profit made on the physical sale. Hedging is an insurance, not a speculation.

It is a valuable protection for all sectors of industry, notably the merchants, whose business could be wiped out by an unpredictable price fluctuation even though the company may be run most efficiently. In other words, it enables the company to get on with its business activity without worrying too much about price fluctuations.

In some respects, therefore, it can be argued that it is speculation to reject the use of hedging where it is possible, since there is a risk in judging the price trends correctly, unless it is a 'back to back' deal under which a matching purchase and sale is made on the same terms.

The Metal Exchange is an unusual commodity futures market. Unlike other futures markets it has an important role in the physical trading of metals. Business on the normal commodity futures market and indeed on Comex is usually confined mainly to 'paper' transactions dealing in future delivery dates and cancelling out any commitment to buy or sell by a matching sale or purchase before the delivery date falls due. Non-trade speculators are not geared to take delivery of a metal and are certainly not able to supply any metal they might have agreed to sell. They must therefore cancel any commitment before the delivery date. The trade also uses futures markets as a means of hedging and price fixing, rather than a source of supply since the metal or commodity deliverable under the contract is often not of the type required by the purchaser. Normally only a small percentage of total turnover is represented by physical trading, the opportunity to make actual purchases and sales is only there as a back-up to the 'paper' trading — a final guarantee or commitment by the user of the market.

The London Metal Exchange

However, the Metal Exchange is an important source of supply for the metals traded, especially copper. It has warehouses dotted throughout Europe — mainly in Britain, Belgium and Holland. These warehouses, usually situated close to the main ports, hold stocks of the brands of metal authorized by the Metal Exchange to be delivered against the contract specifications.

At one stage, producer supply contracts with consumers used to include a clause forbidding the resale of their copper brands on the Metal Exchange. This was to prevent consumers depressing the market by delivering producer copper into the warehouses and then buying it back at a lower price. But this restriction was swept away by the huge build-up of surplus stocks after the 1974 boom. Surplus copper stocks in the warehouses climbed from around a bare 10,000 metric tons during the period of great scarcity in 1974 to over 640,000 metric tons by 1978. This represented a major source of additional supplies, equivalent to the annual production of Zambia at the time. It was available to consumers over and above the normal supplies provided by producers. The Metal Exchange is a market of last resort for both producers and consumers. It is the ultimate dumping ground for any surplus production and is also the alternative source of supplies for consumers when faced with a shortfall, for whatever reason, from their normal suppliers. One major advantage of buying via the Metal Exchange is that the contract does not include any *force majeure* clause, unlike the producer contracts. So the buyer can be sure of obtaining his purchase, whatever happens, unlike producer supplies where deliveries are often cut back as a result of strikes, production disruptions or other dislocations that come under the *force majeure* clause.

For this reason, and to avoid too much dependence on producers, consumers like to buy some of their supplies on a regular basis via the Metal Exchange. This enables them to top up supplies when required and also to sell off surplus holdings to the merchants forming the bulk of the Metal Exchange membership. Ring-dealing members of the Exchange, in fact, incorporate the interests of many of the leading producers, but they have to observe the same contract regulations as the other non-producer controlled members. The Exchange exists mainly for the merchants, who play an influential role in promoting trade across political barriers and ensuring the distribution of metal supplies where they are needed. Since merchants are often at risk — they are forced to take a position rather than rely entirely on back-to-back deals — the hedging facilities provided by the Exchange are crucial to their operations, giving them a much greater flexibility in serving their customers. Producers often use the Exchange's reservoir of supplies to avoid declaring *force majeure* and thereby tempting the customer to move elsewhere. One difficulty, however, is that many customers buy copper from a particular producer because of the quality of the brand which suits their particular needs. The brand bought via the Metal Exchange cannot be guaranteed, except that it meets the quality specifications laid down under the contract. Brands have to be registered with the Metal Exchange by the producer. They are then checked to make sure that the quality specifications of the contract are met before registration is authorized.

Under the Metal Exchange contracts there is a seller's option. This means that the seller has the choice of brand and warehouse from which he will make available the metal purchased by the buyer. However, premiums and discounts are negotiated for different brands, and it is usually possible for the dealer to switch sales and purchases around to the most convenient warehouse for the buyer. Trading is in warrants, specifying the quantity and type of brand, which the buyer presents to the warehouse when he wishes to take delivery. He then arranges transport for the metal from the warehouse to his factory or to another destination — possibly to be transhipped to another buyer. The warrants are the currency used by the Metal Exchange dealers, and are often used as collateral for bank loans or financial guarantees.

The warrants are freely transferable, although their movements have to be reported to the Bank of England to ensure that they are not used as a means of transferring money out of Britain in defiance of the regulations controlling foreign exchange and the outflow of sterling.

Warehouse facilities are strictly regulated by the London Metal Exchange secretariat to ensure that the metal is stored in suitable conditions and that it tallies with the quality and brands specified in the warrants issued by the warehouse managers. Each week the Exchange reports variations in the size of stocks held in the warehouses and this can be an important influence on the market prices.

The Exchange warehouses are the central point at which the residual supplies, not immediately required by consumers, are held and these provide a good barometer of the overall state of the market. In times of surplus the stocks in the warehouses build up and depress the market price. In times of scarcity they dwindle and this forces prices up, often creating a 'backwardation' — that is when the spot or cash price for metal, immediately available from the warehouse, is at a premium to the price of metal for delivery at some time in the future. When the cash price moves to a premium over the three months quotation, it usually means that there is a shortage of supplies worldwide, although on occasions there can be technical squeezes on supplies created by shipping delays to Europe or a sudden surge in demand, which cannot be immediately met. The fundamental principle of the Exchange is that prices go up and down to reflect and eventually correct the supply demand balance. If there is a shortage, prices go up to the level at which any surplus stocks available will come to the Exchange, and producers will be encouraged to step up output. In times of surplus, the build-up stocks in the warehouses depress prices and discourage production whilst stimulating demand. That is the theory although it does not always work in practice.

The Metal Exchange warehouses are amongst the cheapest and easiest places in the world to store surplus metal supplies. The attraction of putting copper and the other metals traded on the Exchange into the warehouses, is that the burden of financing them can be shifted from the producers' or consumers' shoulders. When there are adequate supplies there is normally a 'contango' on the market — that is, the cash price is a discount to the forward quotation; the opposite of backwardation. The gap between the cash price and forward quotation in a contango situation is regulated by the current interest rate in financial markets. This represents the cost of carrying the metal, or holding it in stock, over the period concerned. Banks and other financial institutions help finance the stocks by using what is in effect a guaranteed profit equivalent to or sometimes slightly above the inter-bank interest rate. The procedure is that the bank buys metal at the ruling cash price, and sells at the higher three months price exactly the same quantity of metal for delivery in three months time. It then delivers the metal bought at the lower cash price in three months time against the sale made.

The difference between the lower cash price at which the purchase is made compared with the sale at the higher three months price is the guaranteed profit. Storage and insurance charges for holding the metal at the warehouse (the bank only takes delivery of the warrant) and the brokers' commission have to be deducted. When these deductions from the profit are made they equal the interest rate required to make the proposition sufficiently attractive. This limits the size of any contango, because once it becomes more profitable than the current interest rate, money flows in until it is no longer out of line. Transactions of this kind are known as 'borrowing' from the market.

In a backwardation situation supplies of metal are attracted to the market by the prospect of being able to sell at a higher price and buy back the equivalent amount for delivery at a future date. This is known as 'lending' to the market. The size of the backwardation is theoretically unlimited although it is normally controlled by lending of metal even if there is an overall shortage. Speculators also play a helpful and sometimes vital role in providing financial resources to the market. If a metal is considered to be cheap, speculators are tempted to buy in anticipation of its price going up and generating a useful profit. Speculators tend to prefer buying rather than selling, since it is easier to grasp the concept of buying something one does not want (or of which one does not intend to take delivery) rather than selling something one does not possess. There is a certain logic in this, in that with a sale (or going short as it is known) the sky may be the limit if the price rockets up, especially if one is caught in a technical supply squeeze and has to pay enormous premiums to meet the market commitment by making a matching purchase.

With a purchase (going long) at least the metal bought will always have some value, even if it may be considerably below the price at which it was bought. The attraction of funds from speculators and financial institutions is that they are, in effect, helping to finance surplus stocks, a burden which would normally be carried by producers. At the same time speculative investment in metals, as a safeguard against inflation, currency changes or simply in order to make a profit, means that more money is flowing into the metals industry.

On balance it is true to say that the trade takes more money from speculators, especially at times when it is most needed, than speculators take out of the markets in the form of profits. It is only when the market is artificially distorted by excessive speculation that harm can be done. Over the years the Exchange has learnt to cope with the threat posed by speculators or even by trading companies trying to 'corner' the available supplies and in this way to force up the price. Any sign of this happening is quickly spotted and dealt with by the Metal Exchange management committee, which has the ultimate sanction of withdrawing ring-dealing privileges from any company involved.

It is the trust and confidence built up over the past hundred years that gives the Metal Exchange a good reputation throughout the world, even if many producers and consumers neither care for the pricing system imposed on them nor for the method by which prices are decided.

The Exchange was founded by a group of merchants and traders in London during the 1870s. It formalized previous informal trading sessions where protection was sought against price fluctuations by forward dealings, and which included simply trading in metals. At first one of the members simply drew a large chalk 'ring' on the floor around which other traders gathered to shout out the prices and quantities of metals that they wanted to buy or to sell. Since then a ring of wooden benches with 40 seats for the ring-dealing member companies has replaced the chalk circle. The open cry trading system is used. Dealers shout across the ring during trading sessions the prices at which they wish to buy or sell, and the quantity involved. Deals are struck across the ring between the individual traders, whose clerks take down details for the contracts later required and also keep the company office fully informed of what is happening. This information is then swiftly flashed by modern communication methods throughout the world.

The Exchange has formal trading sessions for the metals traded, spaced out through the day. This enables deals made across the ring to be completed and clients to be kept informed of what is happening so that they can give fresh buying or selling orders before the next trading session starts. In fact the official trading sessions on the ring are confined to the morning. The all-important settlement price, on which many supply contracts are based, is the selling price of cash metal at the end of the morning trading sessions. This is to enable the settlement price to be notified to customers internationally in time for the next day's trading session when they might want to take some further action. Trading then resumes on an unofficial basis in the afternoon, although under the same rules and regulations. The afternoon session coincides with the start of trading in the United States, which as explained earlier can have a considerable influence on prices. In addition to the trading sessions for the individual metals, there are also kerb trading sessions covering all the metals traded. The term 'kerb' dates back to the time when some members, not wishing to participate in the ring trading, held separate dealings on the kerb outside the Exchange. In fact nowadays trading goes on all the time in inter-office dealings between members and their wide network of customers throughout the world. The trading sessions crystallize the inter-office dealings and provide an opportunity to mirror the mood of market sentiment and trends. The good dealer can save or make considerable sums of money for his customers and company by the speed and astuteness with which he handles trading across the ring.

Ring-dealing membership of the Exchange is not easy to obtain. Seats cannot be bought and sold as they are on other Exchanges, since membership of the Exchange is by election only and each company has to apply for re-election annually. A company wishing to become a ring-dealing member has to be able to prove that it can meet the financial commitments entailed, which can be massive, and the Exchange is insistent that the company must be involved in the metals industry rather than simply dealing as brokers on behalf of clients. Election is by secret ballot. In recent years the Exchange has insisted that companies owned by the same parent group should only have one seat on the ring even if they operate independently. They also insist that there must be a majority shareholder bearing the main financial commitment, in any consortium formed to create a dealing company. For much the same reason, any ring-dealing member must be based in London, although this does not preclude overseas companies forming a London subsidiary. The rule that any ring-dealing company should be housed within a mile of the Exchange has been relaxed, although it is a great advantage since each morning, before the Exchange opens for trading, representatives of the member companies sort out details of the previous day's trading, and exchange warrants. The building at Whittington Avenue, where the Exchange has been situated since 1882, is owned by the Metal Market Exchange Company and its Board is responsible for providing the secretariat required to handle the paperwork and ensure that the right facilities for trading are provided.

But the day-to-day running of the market is in the hands of a Management Committee elected from the membership of the Exchange on an annual basis. This committee in turn elects a chairman and vice-chairman. The functions of the management committee and the Board do overlap but the committee's area of responsibility includes trading rules, regulations and policy. This two-tier structure dates back to a dispute in the early days of the Exchange when the Board at the time tried to impose trading regulations which were opposed by a majority of the membership. They then formed the management committee to take over this part of the running of the Exchange. Special sub-committees are formed to deal with specific matters, often in consultation with the Board.

The Bank of England

The Bank of England keeps a watchful eye on the London Metal Exchange as part of its general supervision of the UK commodity markets. It grants a special dispensation to recognized commodity trading companies under schemes devised to allow the international trading of metals and other commodities without going through the normal cumbersome foreign exchange control procedures. In return the Bank insists on detailed reports of all trading so that it can monitor what is happening and ensure that the foreign exchange regulations are not being breached. It has the final sanction through the threat of the withdrawal of the special foreign exchange schemes, without which the market would be unable to operate on an international basis. Since at least 70 per cent of the Metal Exchange turnover comes from outside Britain, it is dependent on the Bank of England for its continued existence. In 1974, the Bank, concerned at talk of excessive speculation, extended its monitoring of the commodity markets to include more information on trading. This was done in order to prevent unhealthy, dangerous situations building up and also to put the brake on speculation when it reached too high a percentage of total turnover.

The Metal Exchange was forced to suspend trading in the zinc market on one occasion in 1976 because of excessive speculation and the Bank has also, on several occasions, issued warnings to other markets.

The Bank put considerable pressure on the Metal Exchange to introduce some form of central clearing facility because of fears that a financial disaster hitting one company could have a domino effect and wreck the entire Exchange.

The 'principals only' system

Another unique feature of the London Metal Exchange is that it has a principals only system. Other commodity futures markets have a central clearing or settlement house through which all transactions are channelled and which takes all the financial risks on behalf of its members. In London the International Commodities Clearing House, a subsidiary of the finance company United Dominions Trust, provides the clearing house for most of the non-metal (soft) commodity futures markets. In the United States the exchanges in New York and Chicago prefer the settlement house system, which is owned by members of the market who thereby spread any financial risk amongst many companies.

On the Metal Exchange each of the ring-dealing members act as principals to any transactions. Trading is not with the Metal Exchange or with a clearing house but with the individual company using a standard contract which complies with rules and regulations laid down by the Metal Exchange. There is also an arbitration procedure in the event of any dispute. This means that each company has to bear the full financial risk on each transaction and, with the growth in trading and the rise in the value of metals, the sums at risk are enormous. Traders attempt to keep a balanced 'book', that is with purchases and sales matched, but this is no real protection against defaulting since both transactions can be at risk: either the purchaser fails to pay what is due or the seller fails to deliver the metal paid for or committed to delivery elsewhere. There is a compensation fund, made up of contributions from the ring-dealing member companies which covers losses that might occur in trading between members. It does not cover any losses that might be sustained by non-members who have to channel their business through a ring-dealing company in order to trade on the Exchange. The compensation fund would not be large enough to cope, even in a limited way, with any sizeable disaster; nor would the cash and bank guarantees put up by ring-dealing members when they are elected to membership.

The 'clearing house' system

There are sharply divided views on the Exchange about the introduction of a clearing house system. Some favour it, others claim that it would change the whole unique concept of the market. With a principals' contract the individual companies can decide the scale of deposits and financial guarantees they require from their clients wishing to do business. Since they do not have to provide deposits to a central clearing house, the ring-dealing members can afford to encourage trade by lowering the deposits they require or indeed by not asking for any deposit provided the customer is financially sound and not over-committed. In other instances, notably with speculative clients, the amount required as a deposit on forward transactions is normally at least 10 per cent of the total outlay and sometimes considerably more, if the client is thought to be financially unreliable.

Dealing ahead

At the same time the principals' contract is linked with another unique feature of the London Metal Exchange: the fact that it is possible to trade on any business day up to three months ahead on the Exchange and considerably further ahead with individual members. The official quotations on the Exchange only extend up to three months ahead, unlike other futures contracts where trading is possible in the specified delivery months a year ahead or in some cases up to three years ahead. The three months period on the Exchange dates back to the time when it took that time for copper and Straits tin to reach London from Chile and Malaya respectively.

It has not been extended because of the cost involved in members granting credit over a longer period plus the fact that it would be difficult to extend the daily delivery dates, which are much appreciated by trade customers, for longer than three months. However, there is no difficulty in dealing further ahead with any ring-dealing member. The current quotation is simply added to the interest rate involved in carrying the metal required for a longer period and a forward price is calculated.

Under pressure from the Bank of England the Metal Exchange decided in 1978 to proceed with plans for a monitoring system, aimed at giving a much clearer picture of the individual members companies' commitments. This would allow for action to be taken if a dangerous situation appeared to be building up, and the guarantee sums required could be adjusted according to the volume of business and the risk undertaken by each member company.

Copper and Tin

Copper is by far the most important market on the Metal Exchange, which is often described as the London Copper Exchange. It has more trading sessions, since there are separate sessions for wirebars and cathodes — the two contracts that have finally evolved. But tin has also been traded on the Exchange since its inception, although obviously in a lower volume as it is far less plentiful than copper. Since the advent of the International Tin Agreement which followed on previous attempts to control the market, the Metal Exchange has played a less important role in influencing tin prices. This role diminished still further when the International Tin Council decided to base its buffer stock ranges on the Straits tin price quoted on the Penang market in Malaysia because of the float of sterling in 1972.

In Britain the collapse of the Williams Harvey smelter in 1973, which had been a major source of tin supplies to the Metal Exchange, brought a further loss of influence, despite the introduction of a new high-grade contract which is geared to allow for trading in the bulk of tin supplies throughout the world. At the time of writing, in 1978, the Metal Exchange market was the only hedging medium available for tin traders throughout the world and thus still plays an important though diminished role. It may be argued that with the International Tin Agreement controlling prices there is less need for hedging facilities but this is not the case. The range in which tin prices can, and do, move under the Agreement is wide enough to create severe losses for the trader miscalculating the trend in prices. In some ways the Tin Agreement acts as a destabilizing influence since the limits set out under the Agreement provide a target for speculators, both trade and private, to shoot at in the knowledge that specific action will be triggered off — often providing the opportunity for making profits.

At the same time, although the floor price of the Agreement has rarely been breached, there is still only a partial defence against prices going through the ceiling in times of shortage and the buffer stock losing control of the market. Indeed this is recognized, somewhat unfairly in the Tin Agreement, in that while the buffer stock resources are used to defend the floor price, the sale takes place at the market price, possibly above the ceiling when the top limit is breached. At the same time the dependence of Europe and the LME warehouses on supplies shipped from some distance away means that there is far greater vulnerability to supply 'squeezes' through the dislocation of deliveries.

The amount of tin produced in the main industrialized countries is limited, with the production of the actual tin metal centred around the areas of the mine output of concentrates. This is different from copper, lead and zinc, where a large proportion of the metal is produced in the consuming countries.

The most important physical market for tin is in Penang, where the world's two major tin smelting companies offer their production on a daily basis against bids from buyers. The offers are based on the intake of concentrates from the mines by the smelters. Each day the smelters compile a list of bids received against the supplies available. Starting from the highest bid they work down until supplies for that day are exhausted and the final price is the price for the day. Many traders, of course, hedge their Penang purchases on the London Metal

Exchange and there is a close link between the two markets.

In 1977 the Malaysian Government commissioned a study on the prospects for starting a tin futures market, for forward dealing, to operate in conjunction with the Penang physical market. It is part of an overall policy to retain in Malaysia some of the control and profits resulting from futures trading in tin and several other of its main commodity exports. It can be expected, however, that the Metal Exchange market, with its long history and expertise, will retain its important role, especially since it is centred in a consuming area and not as subject to producer influences.

Lead and Zinc

The Metal Exchange lead and zinc markets do not have a long history. They were introduced for the first time in 1920 and like the other markets, were closed during the Second World War and reopened only in 1953.

The lead market is very similar to copper, except that it operates on a smaller scale and is far less active thus reflecting the difference between the two metals. Lead has a history of relative price stability since its consumption tends to grow slowly and steadily unlike the violent changes from boom to bust in other metals. A dampening influence on market price movements is the fact that its major outlet, in lead-acid batteries, generates a large proportion of new supplies in the form of recovered scrap. Secondary or recovered lead can be refined back into virtually the same quality as primary lead, and there are few uses that require top purity lead. In Britain and other industrialized countries, where the system for recovery of lead, especially from batteries, has reached a high degree of efficiency, as much as 60 per cent of total supplies comes from scrap sources. The comparable figure for copper is some 40 per cent and it is far less for most other metals, with the exception of aluminium.

There have been occasions when lead prices have shot up as a result of a sudden surge in demand, or more often as result of the dislocation of supplies in the main producing areas in North America and Australia. But the flow of scrap supplies usually can be rapidly increased given sufficient price incentive.

As with copper, the bulk of lead is sold to consumers under supply contracts with producers, using the Metal Exchange quotations as the basis for the price paid. The Metal Exchange provides the basis for costs of concentrates and ores bought by smelters, as well as for scrap supplies and manufactured metal. It is also an important alternative source of physical supplies, again, like copper, acting as a residual market of last resort.

Lead producers, who are mainly centred in developed countries, have over the years attempted to influence the Metal Exchange prices to avoid the kind of fluctuations encountered in copper. Since lead is a smaller market, it is far easier and less costly to undertake support-buying operations. The purchase of a relatively small quantity of lead on the Exchange can be more than justified if, as a result, the price of the bulk of lead sold under supply contracts direct to consumers is increased. Generally the producers' aim is to lift prices, or at least to prevent them from falling. On occasions the selling of surplus metal is used to stop too violent a rise in price which it is feared might hit consumption.

However, despite the smaller scale of operations needed to control the market, on occasions producers find it impossible to withstand normal market forces pushing prices down.

These downward pressures were naturally at their strongest during times of surplus supplies, when producing companies were at their weakest with sales falling and their own stocks building up. The cost of carrying extra supplies, acquired under support-buying operations from the Metal Exchange, put producers under a severe financial strain. The alternatives were to bow to pressure and allow prices to fall or to be stuck with surplus stocks acquired at a higher price. The large percentage of scrap supplies in the market, from suppliers with totally different interests and cost structures from primary producers, was a flaw in the whole conception. But the break-up of combined operations by a group of leading producers sharing the cost came when the United States and other governments showed their hostility to this kind of cooperative action, which, it was claimed, ran counter to anti-cartel legislation. The United States Government had a particular interest since the price charged by domestic producers under the producer price system has generally to be aligned with the level of prices in Europe and elsewhere, allowing for shipping costs and the import duty imposed as a protection for US producers. At the same time the main sources of lead production are North America and Australia, where anti-trust legislation has been vigorously enforced. Producers became increasingly nervous about the support-buying operations as it became clear that many companies could not cooperate for fear of being charged with anti-trust activities. In addition the cost of trying to arrest the general decline in prices during the economic recession became prohibitive. Support-buying operations were therefore suspended, probably indefinitely. This did not include individual producers, although their existence has never been officially acknowledged.

Similar problems have faced the zinc producers' group that in 1964 decided to abandon the use of the Metal Exchange quotations as a pricing basis for supply contracts. Instead, they instituted what is known as the European producer price, used as the basis for contracts throughout the world outside North America which has its own producer price system. In order to maintain the European producer price support-buying operations were needed to take up surplus supplies on the London Metal Exchange, when prices had been driven down to a level which might tempt consumers away from producer supplies to the cheaper LME zinc. This was not a particularly difficult operation since the supplies available to the Metal Exchange zinc market were limited mainly to shipments from Communist bloc countries, notably North Korea. Only in times of gross surplus could the Metal Exchange rely on adequate supplies being available, as a result of reselling by consumers, or from producers desperate to sell their surplus zinc. Merchant interest in zinc declined as a result of the producers grip on the market and trading activity in the free market followed suit with only occasional bursts of interest. The weakness of the LME zinc market was starkly revealed in 1973 when a shortage of available supplies forced the cash price up to an unprecedented £950 per ton − about three times the producer price at the time − and the Metal Exchange had to suspend the selling of cash metal for a period, until more normal conditions were restored.

The European zinc producer price is, in theory, easier to sustain than the producer price for copper and lead, since there is no comparable competitive pressure from scrap supplies. Control of production is centred in relatively few hands, mainly in developed countries, where producers are better able to adjust production in accordance with the changing demand. These producers normally have the facility and financial resources to cut output when necessary, unlike copper producers in developing countries, who cannot afford either the loss of export earnings or the social problems created by increased unemployment.

In fact, the European zinc price, instituted on the initiative of the giant Rio Tinto Zinc group and strongly backed by Belgian and French producers, established itself successfully and worked to the satisfaction of both producers and consumers for over 10 years, with occasional slight hiccups. Meetings between producers were always held unofficially, but decisions on production cutbacks and price increases were quickly made known. Nowadays producers using producer price systems prefer to use a follow-my-leader approach. One company raises or cuts its producer price, other producers decide individually whether they should follow or not, and a general consensus emerges.

However, moves for the stricter enforcement of anti-trust legislation coincided with a period when the support-buying operation was facing a crisis as a result of a slump in demand. This created massive surplus stocks. Producers were in disarray, fearful of meeting together even unofficially, and having to cope with heavy extra costs to finance stocks at unprecedentedly high levels.

The producer price, which had been quoted in sterling changed to US dollars at the end of 1974. It had been held too high for too long and was heavily discounted by producers and smelters battling for a share of declining sales. Eventually pressure from smelters forced a reduction in the producer price from $795 to $550 a ton in less than two years. An underlying problem for the zinc producers is that smelters, not owned by the mining companies, have a similar attitude to scrap refiners. They are most interested in the difference between the price at which they buy their raw material intake of ore concentrates and the price they can obtain for the refined metal, which represents their profit margin.

This conflict between the interests of mining producers, many of whom also own zinc smelting plants, and the independent European smelters is a constant threat to the European zinc producer price structure. It has continued to exist only because the stability is favoured by consumers and is used as the basis for trading in ore concentrates worldwide. Canadian producers, who provide the bulk of the world's zinc exports, are keen supporters of the producer price system. The big cutback in US zinc refining capacity in the late 1960s which was encouraged by environmental problems now means that the United States has become a large net importer. US buyers historically favour a producer price system. Import competition, however, means that US domestic prices and international prices are even more closely linked than in the past.

The Metal Exchange market, handicapped by the lack of a reliable, regular source of supply, faces a difficult task in operating effectively. The serious blows suffered by the European producer price system have established the Exchange market in a much more influential role in setting price trends. A zinc futures contract was established on the New York commodity exchange in 1978 to cater for the increased flow of imported supplies. The initial reaction to this, as with the reaction to other base metal markets in the United States, with the exception of the Comex copper market, was not encouraging.

Sample LME contract

Authorised 25th July, 1978

PRIMARY ALUMINIUM CONTRACT FORM

**Approved by the Board of Directors and by the
Committee of the London Metal Exchange**

Contract B

LONDON, ...

M...

$\dfrac{\text{I}}{\text{We}}$ have this day $\dfrac{\text{sold to}}{\text{bought from}}$ you, according and subject to the Rules and

Regulations of the London Metal Exchange,

....

.... TONNES (two per cent. either more or less and
subject to Rule 3 below) of

PRIMARY ALUMINIUM

Price £.. per tonne of 1000 kilogrammes net $\dfrac{\text{plus}}{\text{minus}}$ % to us

....

The appropriate Import Duty (if any) ruling on the prompt date to be for Buyer's Account.

Prompt ....

We have the right at any time on demand to require you to pay us such a sum (hereinafter referred to as a "margin") in cash and/or to deposit with us security in such other form and of such amount not exceeding the value of the contract as we in our discretion require and in order to secure the due fulfilment by you of your obligations under this contract and to the intent that the value of the margin in relation to the contract shall at all times during the currency of the contract be maintained by you we have the further right on demand and whether in one or more calls to require you to pay to us the difference between the value of the contract at the time of entering into the same and the current market value at any time thereafter as we in our discretion require. In the event of any failure by you to fulfil your obligations we have an immediate right of appropriation of any such cash and/or to sell any security to satisfy our rights as above in addition to all other rights reserved to us by this contract.

This contract is made between ourselves and yourselves as principals, we alone being liable to you for its performance. The percentage (if any) charged by us to you is to be regarded simply as part of the price and may be shared by us with agents introducing the business, whilst we reserve the right also to charge a percentage or commission to any person from or to whom we may have bought or sold to cover our liability hereunder.

In the event of your failing to meet your engagements arising out of this or any other outstanding contract or contracts between us which are subject to the Rules and Regulations of the London Metal Exchange whether by failing to provide on the due date documents to meet sales or money to take up documents (as the case may be) or otherwise howsoever or of your failing to supply or maintain such margin (if any) for which we are entitled to call and have called or in the event of your suspending payment or becoming bankrupt or committing any act of bankruptcy or (being a Company) in the event of your going into liquidation whether voluntary or otherwise, we reserve the right to close this contract and any other said outstanding contract or contracts if as and when we in our sole discretion shall so decide by selling out or buying in against you (as the case may be) and any differences arising therefrom shall be payable forthwith notwithstanding that the prompt day or other day originally stipulated for settlement may not have arrived.

Any delay in our enforcing any of our rights under this contract shall not be deemed to constitute a waiver thereof.

Members of the London Metal Exchange

The same contract is used for copper cathodes, copper wirebar, standard lead, high grade tin, standard tin and standard zinc.

SPECIAL RULES FOR PRIMARY ALUMINIUM

1. Quality.—The Aluminium delivered under this contract shall be Primary Aluminium of minimum 99·50% purity, with maximum Iron content 0·4% and maximum Silicon content 0·30% delivered in the form of ingots securely strapped in bundles suitable for stacking. All Aluminium delivered must be of the brands listed in the L.M.E. approved list of Aluminium brands.

2. Settlement.—Contracts shall be settled on exact quantities of 25 tonnes at the official settlement price quoted by the Committee operative on the prompt date, buyer and seller paying or receiving, as the case may be, the difference if any between the settlement price and the contract price.

3. Delivery.—The Aluminium shall be delivered on the prompt date in any warehouse listed for this purpose with the Committee at listed delivery points in seller's option. (Names of such warehouses can be obtained from the London Metal Exchange Office.) Warrants tendered in fulfilment of contracts shall be invoiced at the settlement price mentioned in Rule 2 above in parcels each of 25 tonnes or a multiple thereof (each 25 tonnes to be treated as a separate contract). Warrants shall be for 25 tonnes each (two per cent. either more or less). Each parcel of 25 tonnes shall be the production of one country, shall consist of ingots of one brand, shape and size and shall lie at one warehouse. Rent shall be allowed on the invoice.

4. Weights.—Warrant weights in all cases shall be accepted as between buyer and seller. The weight of each ingot shall be within the range of 12kgs. to 26kgs.

5. Placing Aluminium on Warrant.—(a) Each delivery of Aluminium must be accompanied by a Certificate of Origin and a producer's Certificate of Analysis, both of which documents must be lodged with the Warehouse-man. The Analysis Certificates must be within the prescribed quality specification. A Bulk Analysis Certificate or a copy thereof is acceptable, but if a producer's Certificate is not available, an Analysis Certificate must be prepared and signed by a London Metal Exchange listed assayer. All Analysis Certificates must show each heat number.

(b) The party intending to place Aluminium on warrant for delivery to the Market must advise the Ware-houseman the name of a Ring Member who will be responsible for the initial placing of the warrant(s) on the Market. Such warrant(s) must be issued to the order of that Ring Member who prior to endorsement of the warrant(s) must be satisfied that the documentation is evidence of good delivery.

6. Warrants.—(a) Each warehouse warrant shall be made up of Aluminium from the production of one country and shall be of brands as listed on the London Metal Exchange approved list of Aluminium brands.

The warrant shall indicate the country of origin, brand and the date(s) and reference number(s) of the Certifi-cates of Origin and of Analysis lodged with the Warehouseman. Each warrant shall be of 25 tonnes (2 per cent. either more or less), shall lie in one warehouse and shall indicate the number of bundles and the number of ingots comprising each bundle.

(b) On request the Warehouseman is obliged to submit to the holder of the warrant, Certificates of Origin,/ Analysis or copies if the 25 tonnes form part of a larger delivery covered by bulk analysis.

7. Exchange Control Regulations.—In the event of a resident of the United Kingdom taking delivery of warrants in London Metal Exchange listed warehouses outside the United Kingdom, the buyer must conform with the Bank of England Metals Scheme C.M. procedure currently in force.

8. Disputes.—Any question concerning formation and any dispute under this contract shall be notified to the Executive Secretary to the Committee of the London Metal Exchange in writing by the seller or buyer or both of them jointly. Such question or dispute if not settled by agreement shall be referred to arbitration in accordance with the Rules and Regulations of the London Metal Exchange. The decision in writing of the Appeal Committee from the Award of the Arbitrators shall be a condition precedent to a Notice of Motion to remit or set aside the Award of the Arbitrators and the decision of the Appeal Committee or to an action being brought. The Uniform Law concerning the formation of contracts for the International Sale of Goods and the Uniform Law regulating the International Sale of Goods shall not apply.

(In the above Rules "The Committee" means the Committee of the London Metal Exchange.)

For Contract Rules see over.

SPECIAL RULES FOR COPPER CATHODES

1. Quality.—The copper delivered under this contract must be Electrolytic Copper in the form of cathodes assaying not less than 99·90 per cent. of copper (*silver being counted as copper*).

All Copper delivered must be of brands approved by and registered with the Committee, and must conform with the current appropriate standard of either the B.S.I. or the A.S.T.M.

2. Settlement.—Contracts shall be settled on exact quantities of 25 tons at the official Settlement price quoted by the Committee operative on the prompt date, Buyer and Seller paying or receiving, as the case may be, the difference, if any, between the Settlement price and the contract price.

3. Delivery.—The Copper shall be delivered on the prompt date in warehouse; either London, Birmingham, Manchester, Liverpool, Birkenhead, Hull, Newcastle-on-Tyne, Glasgow, Avonmouth, Swansea, Rotterdam, Bremen, Hamburg (free port area) or Antwerp in seller's option. In all cases the warehouse must be one approved by and registered with the Committee. Warrants tendered in fulfilment of contracts shall be invoiced at the Settlement price mentioned in Rule 2 above in parcels each of 25 tons or a multiple thereof (each 25 tons to be treated as a separate contract). Warrants shall be for 25 tons each (two per cent either more or less). Warrants issued prior to the 1st January, 1970 for long tons shall constitute good delivery provided that their weights are within 2 per cent tolerance of 25 long tons. Each parcel of 25 tons shall be of one brand, shape and size of full or cut plates subject to the necessity of including different shapes and sizes at the bottom of each parcel for the purpose of palletisation, and shall lie at one warehouse. Cut plates shall not be less than quarter size. Rent shall be allowed on the invoice.

4. Weights.—The word "TON" wherever appearing in this contract shall be a metric tonne of 1000 kilogrammes. In the case of warrants where weights are shown in long tons conversion shall be at the rate of 1 long ton to 1016 kilogrammes. Warrant weights in all cases shall be accepted as between buyer and seller.

5. Warrants.—Each warrant must state the name of the producer and the number of plates comprising each parcel.

6. Exchange Control Regulations.—If Rotterdam, Bremen, Hamburg or Antwerp warrants are delivered to a resident of the United Kingdom not participating in the Bank of England Metals Scheme in fulfilment of this contract, the seller must issue a C.M. form in accordance with the procedure currently in force under the Scheme. The buyer must conform with the requirements expressed on those parts of the C.M. form which he receives from the Seller.

7. Disputes.—Any question concerning formation and any dispute under this contract shall be notified to the Secretary of the London Metal Exchange in writing by the seller or the buyer or both of them jointly. Such question or dispute if not settled by agreement shall be referred to arbitration in accordance with the Rules and Regulations of the London Metal Exchange. The decision in writing of the Appeal Committee from the Award of the Arbitrators shall be a condition precedent to a Notice of Motion to remit or set aside the Award of the Arbitrators and the decision of the Appeal Committee or to an action being brought. The Uniform Law concerning the formation of contracts for the International Sale of Goods and the Uniform Law regulating the International Sale of Goods shall not apply.

(In the above Rules "The Committee" means the Committee of the London Metal Exchange)

For Contract Rules see over

SPECIAL RULES FOR COPPER WIREBARS

1. Quality.—The Copper delivered under this contract must be in the form of wirebars of standard dimensions in the weight range 90 kgs to 125 kgs, and may be either:—

(a) Electrolytic Copper

or

(b) High Conductivity Fire Refined Copper

In the latter case (b) a deduction of £20 per ton shall be allowed on the invoice.

All Copper delivered must be of brands approved by and registered with the Committee, and must conform with the current appropriate standard of either the B.S.I. or the A.S.T.M.

2. Settlement.—Contracts shall be settled on exact quantities of 25 tons at the official Settlement price quoted by the Committee operative on the prompt date, Buyer and Seller paying or receiving, as the case may be, the difference, if any, between the Settlement price and the contract price.

3. Delivery.—The Copper shall be delivered on the prompt date in any warehouse listed for this purpose with the Committee at listed delivery points in seller's option (Names of such warehouses can be obtained from the London Metal Exchange Office). Warrants tendered in fulfilment of contracts shall be invoiced at the Settlement price mentioned in Rule 2 above in parcels each of 25 tons or a multiple thereof (each 25 tons to be treated as a separate contract). Warrants shall be for 25 tons each (two per cent. either more or less). Warrants issued prior to the 1st January, 1970 for long tons shall constitute good delivery provided that their weights are within a 2% tolerance of 25 long tons. Each parcel of 25 tons shall be of one brand, shape and size, and shall lie at one warehouse. Rent shall be allowed on the invoice.

4. Weights.—The word "TON" wherever appearing in this contract shall be a metric tonne of 1000 kilogrammes. In the case of warrants where weights are shown in long tons conversion shall be at the rate of 1 long ton to 1016 kilogrammes. Warrant weights in all cases shall be accepted as between buyer and seller.

5. Warrants.—Each Warrant must state the brand and whether electrolytic or H.C.F.R. and the wirebars comprising each parcel of 25 tons shall be of uniform weight subject to the usual tolerances, and the wirebar weight and the number of wirebars comprising each parcel must be shown on the Warrant.

6. Exchange Control Regulations.—If Rotterdam, Bremen, Hamburg or Antwerp warrants are delivered to a resident of the United Kingdom not participating in the Bank of England Metals Scheme in fulfilment of this contract, the seller must issue a C.M. form in accordance with the procedure currently in force under the Scheme. The buyer must conform with the requirements expressed on those parts of the C.M. form which he receives from the Seller.

7. Disputes.—Any question concerning formation and any dispute under this contract shall be notified to the Secretary of the London Metal Exchange in writing by the seller or the buyer or both of them jointly. Such question or dispute if not settled by agreement shall be referred to arbitration in accordance with the Rules and Regulations of the London Metal Exchange. The decision in writing of the Appeal Committee from the Award of the Arbitrators shall be a condition precedent to a Notice of Motion to remit or set aside the Award of the Arbitrators and the decision of the Appeal Committee or to an action being brought. The Uniform Law concerning the formation of contracts for the International Sale of Goods and the Uniform Law regulating the International Sale of Goods shall not apply.

(In the above Rules "The Committee" means the Committee of the London Metal Exchange.)

SPECIAL RULES FOR STANDARD LEAD

1. Quality.—The Lead delivered under this contract must be Refined Pig Lead (minimum 99·97 per cent. purity).

All Lead delivered must be:—

(*i*) of brands approved by and registered with the Committee.

(*ii*) in Pigs weighing not more than 55 kgs each.

2. Settlement.—Contracts shall be settled on exact quantities of 25 tonnes at the official Settlement price quoted by the Committee operative on the prompt date, Buyer and Seller paying or receiving, as the case may be, the difference, if any, between the Settlement price and the contract price.

3. Delivery.—Warrants tendered in fulfilment of contracts shall be invoiced at the Settlement price mentioned in Rule 2 above in parcels each of 25 tonnes or a multiple thereof (each 25 tonnes to be treated as a separate contract). Warrants shall be for 25 tonnes each (two per cent. either more or less). Warrants issued prior to the 1st January, 1970 for long tons shall constitute good delivery provided that their weights are within a 2% tolerance of 25 long tons. Each parcel of 25 tonnes and/or 25 long tons shall be of one brand and shall consist of pigs of one size, subject to the necessity of including different shapes and sizes at the bottom of each parcel for the purpose of palletisation, shall lie at one warehouse and the number of pigs comprising each parcel must be shown on the warrant. Rent shall be allowed on the invoice.

The Lead shall be delivered on the prompt date in any warehouse approved by and registered for Lead with the Committee at delivery points approved by the Board and Committee in seller's option.

4. Weights.—In the case of warrants where weights are shown in long tons conversion shall be at the rate of 1 long ton to 1016 kilogrammes. Warrant weights in all cases shall be accepted as between buyer and seller.

5. Disputes.—Any question concerning formation and any dispute under this contract shall be notified to the Secretary of the London Metal Exchange in writing by the seller or the buyer or both of them jointly. Such question or dispute if not settled by agreement shall be referred to arbitration in accordance with the Rules and Regulations of the London Metal Exchange. The decision in writing of the Appeal Committee from the Award of the Arbitrators shall be a condition precedent to a Notice of Motion to remit or set aside the Award of the Arbitrators and the decision of the Appeal Committee or to an action being brought. The Uniform Law concerning the formation of contracts for the International Sale of Goods and the Uniform Law regulating the International Sale of Goods shall not apply.

6. Exchange Control Regulations.—In the event of a resident of the United Kingdom taking delivery of warrants in L.M.E. registered warehouse outside the United Kingdom, the buyer must conform with the Bank of England Metals Scheme C.M. Procedure currently in force.

(In the above Rules "The Committee" means the Committee of the London Metal Exchange.)

For Contract Rules see over.

Note: **List of approved warehouses can be obtained from the London Metal Exchange Office.**

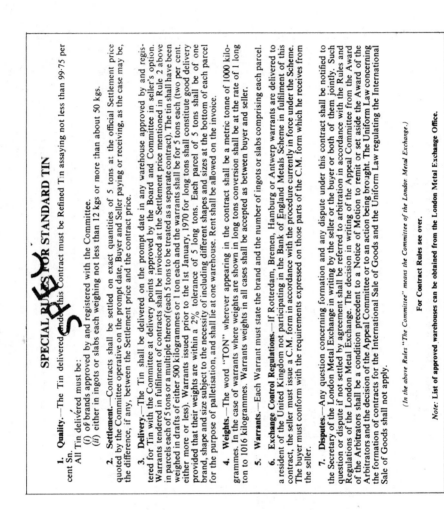

SPECIAL RULES FOR STANDARD TIN

1. **Quality.**—The Tin delivered under this Contract must be Refined Tin assaying not less than 99·75 per cent Sn.

All Tin delivered must be:

(i) of brands approved by and registered with the Committee.

(ii) either in ingots or slabs each weighing not less than 12 kgs or more than about 50 kgs.

2. **Settlement.**—Contracts shall be settled on exact quantities of 5 tons at the official Settlement price quoted by the Committee operative on the prompt date, Buyer and Seller paying or receiving, as the case may be, the difference, if any, between the Settlement price and the contract price.

3. **Delivery.**—The Tin shall be delivered on the prompt date in any warehouse approved by and registered for Tin with the Committee at delivery points approved by the Board and Committee in seller's option. Warrants tendered in fulfilment of contracts shall be invoiced at the Settlement price mentioned in Rule 2 above in parcels each of 5 tons or a multiple thereof (each 5 tons to be treated as a separate contract). The tin shall have been weighed in drafts of either 500 kilogrammes or 1 ton each and the warrants shall be for 5 tons each (two per cent. either more or less). Warrants issued prior to the 1st January, 1970 for long tons shall constitute good delivery provided that their weights are within a 2% tolerance of 5 long tons. Each parcel of 5 tons shall be of one brand, shape and size subject to the necessity of including different shapes and sizes at the bottom of each parcel for the purpose of palletisation, and shall lie at one warehouse. Rent shall be allowed on the invoice.

4. **Weights.**—The word "TON" wherever appearing in the contract shall be a metric tonne of 1000 kilogrammes. In the case of warrants where weights are shown in long tons conversion shall be at the rate of 1 long ton to 1016 kilogrammes. Warrants weights in all cases shall be accepted as between buyer and seller.

5. **Warrants.**—Each Warrant must state the brand and the number of ingots or slabs comprising each parcel.

6. **Exchange Control Regulations.**—If Rotterdam, Bremen, Hamburg or Antwerp warrants are delivered to a resident of the United Kingdom not participating in the Bank of England Metals Scheme in fulfilment of this contract, the seller must issue a C.M. form in accordance with the procedure currently in force under the Scheme. The buyer must conform with the requirements expressed on those parts of the C.M. form which he receives from the seller.

7. **Disputes.**—Any question concerning formation and any dispute under this contract shall be notified to the Secretary of the London Metal Exchange in writing by the seller or the buyer or both of them jointly. Such question or dispute if not settled by agreement shall be referred to arbitration in accordance with the Rules and Regulations of the London Metal Exchange. The decision in writing of the Appeal Committee from the Award of the Arbitrators shall be a condition precedent to a Notice of Motion to remit or set aside the Award of the Arbitrators and the decision of the Appeal Committee or to an action being brought. The Uniform Law concerning the formation of contracts for the International Sale of Goods and the Uniform Law regulating the International Sale of Goods shall not apply.

(In the above Rules "The Committee" means the Committee of the London Metal Exchange.)

For Contract Rules see over.

Note: **List of approved warehouses can be obtained from the London Metal Exchange Office.**

SPECIAL RULES FOR HIGH GRADE TIN

1. **Quality.**—The tin delivered under this Contract must be Refined Tin assaying not less than 99.85 per cent Sn.

All Tin delivered must be:

(i) of brands approved by and registered with the Committee

(ii) either in ingots or slabs each weighing not less than 12 kgs or more than about 50 kgs.

2. **Settlement.**—Contracts shall be settled on exact quantities of 5 tonnes at the official Settlement price quoted by the Committee operative on the prompt date, Buyer and Seller paying or receiving, as the case may be, the difference, if any, between the Settlement price and the contract price.

3. **Delivery.**—The Tin shall be delivered on the prompt date in warehouse; either London, Birmingham, Manchester, Liverpool, Birkenhead, Hull, Newcastle-on-Tyne, Glasgow, Avonmouth, Swansea, Rotterdam, Hamburg (free port area) or Antwerp in seller's option. In all cases the warehouse must be one approved by and registered with the Committee. Warrants tendered in fulfilment of contracts shall be invoiced at the Settlement price mentioned in Rule 2 above in parcels each of 5 tonnes or a multiple thereof (each 5 tonnes to be treated as a separate contract). The tin shall have been weighed in drafts of either 500 kilogrammes or 1 tonne each and the warrants shall be for 5 tonnes each (two per cent either more or less). Warrants issued prior to the 1st January, 1970 for long tons shall constitute good delivery provided that their weights are within a 2% tolerance of 5 long tons. Each parcel of 5 tonnes shall be of one brand, shape and size subject to the necessity of including different shapes and sizes at the bottom of each parcel for the purpose of palletisation, and shall lie at one warehouse. Rent shall be allowed on the invoice.

4. **Weights.**—The word "TONNE" wherever appearing in the contract shall be a metric ton of 1000 kilogrammes. In the case of warrants where weights are shown in long tons conversion shall be at the rate of 1 long ton to 1016 kilogrammes. Warrant weights in all cases shall be accepted as between buyer and seller.

5. **Warrants.**—Each Warrant must state the brand and the number of ingots or slabs comprising each parcel.

6. **Exchange Control Regulations.**—If Rotterdam, Hamburg or Antwerp warrants are delivered to a resident of the United Kingdom not participating in the Bank of England Metals Scheme in fulfilment of this contract, the seller must issue a C.M. form in accordance with the procedure currently in force under the Scheme. The buyer must conform with the requirements expressed on those parts of the C.M. form which he receives from the Seller.

7. **Disputes.**—Any question concerning formation and any dispute under this contract shall be notified to the Secretary of the London Metal Exchange in writing by the seller or the buyer or both of them jointly. Such question or dispute if not settled by agreement shall be referred to arbitration in accordance with the Rules and Regulations of the London Metal Exchange. The decision in writing of the Appeal Committee from the Award of the Arbitrators shall be a condition precedent to a Notice of Motion to remit or set aside the Award of the Arbitrators and the decision of the Appeal Committee or to an action being brought. The Uniform Law concerning the formation of contracts for the International Sale of Goods and the Uniform Law regulating the International Sale of Goods shall not apply.

(In the above Rules "The Committee" means the Committee of the London Metal Exchange.)

For Contract Rules see over

SPECIAL RULES FOR STANDARD ZINC

1. Quality.—The Zinc delivered under this contract must be produced by distillation or electrolysis, and must be Zinc of minimum 98 per cent. purity.

All Zinc delivered must be:—

(i) of brands approved by and registered with the Committee.

(ii) either in Slabs or in Plates or in Ingots weighing not more than 55 kgs each.

2. Settlement.—Contracts shall be settled on exact quantities of 25 tonnes at the official Settlement price quoted by the Committee operative on the prompt date, Buyer and Seller paying or receiving, as the case may be, the difference, if any, between the Settlement price and the contract price.

3. Delivery.—Warrants tendered in fulfilment of contracts shall be invoiced at the Settlement price mentioned in Rule 2 above in parcels each of 25 tonnes of a multiple thereof (each 25 tonnes to be treated as a separate contract). Warrants shall be for 25 tonnes each (two per cent. either more or less). Warrants issued prior to the 1st January, 1970 for long tons shall constitute good delivery provided that their weights are within a 2% tolerance of 25 long tons. Each parcel of 25 tonnes and/or 25 long tons shall be of one brand and shall consist of either slabs, plates or ingots of one size subject to the necessity of including different shapes and sizes at the bottom of each parcel for the purpose of palletisation, shall lie at one warehouse and the number of slabs, plates or ingots comprising each parcel must be shown on the warrant. Rent shall be allowed on the invoice.

The Zinc shall be delivered on the prompt date in any warehouse approved by and registered for Zinc with the Committee at delivery points approved by the Board and Committee in seller's option.

4. Weights.—In the case of warrants where weights are shown in long tons conversion shall be at the rate of 1 long ton to 1016 kilogrammes. Warrants weights in all cases shall be accepted as between buyer and seller.

5. Disputes.—Any question concerning formation and any dispute under this contract shall be notified to the Secretary of the London Metal Exchange in writing by the seller or the buyer or both of them jointly. Such question or dispute if not settled by agreement shall be referred to arbitration in accordance with the Rules and Regulations of the London Metal Exchange. The decision in writing of the Appeal Committee from the Award of the Arbitrators shall be a condition precedent to a Notice of Motion to remit or set aside the Award of the Arbitrators and the decision of the Appeal Committee or to an action being brought. The Uniform Law concerning the formation of contracts for the International Sale of Goods and the Uniform Law regulating the International Sale of Goods shall not apply.

6. Exchange Control Regulations.—In the event of a resident of the United Kingdom taking delivery of warrants in L.M.E. registered warehouse outside the United Kingdom, the buyer must conform with the Bank of England Metals Scheme C.M. Procedure currently in force.

(In the above Rules "The Committee" means the Committee of the London Metal Exchange.)

For Contract Rules see over.

Note: **List of approved warehouses can be obtained from the London Metal Exchange Office.**

Part 2:
The Metals

Aluminium

Aluminium production 1976 *metric tons x 1 million*

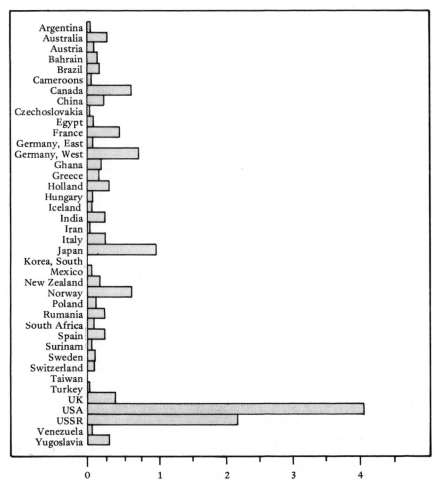

Total world production 12,736,000 metric tons

Aluminium primary consumption 1976

metric tons x 1 million

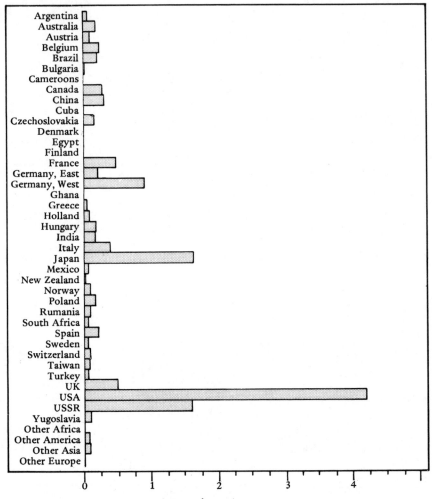

Total world consumption 13,312,600 metric tons

Grades available

The bulk of unwrought aluminium is produced in the form of ingots weighing approximately 22 kilos with the purity of 99.5 per cent minimum or 99.7 per cent minimum. Many smelters, however, can offer a large range of shapes and grades, including 'T' bars — up to 1000 kilos each; sows, 200-500 kilos each; extrusion billets; rolling slabs and wire rods.

Most modern smelters are designed to produce metal with the minimum purity of 99.7 per cent with iron content below 0.2 per cent and the silicon content below 0.1 per cent.

Production method

The metal is extracted from alumina (aluminium oxide) which is produced from bauxite, a naturally occurring ore.

Aluminium is the most abundant metallic element in the earth's crust.

It is produced by the electrolysis of alumina with cryolite, a natural mineral that can also be made synthetically. The anode of the electrolytic production cell is made of carbon which combines with the oxygen in the alumina when the current is passed through and sufficient heat is generated, reducing the alumina to the metal. The molten metal is then cast into the moulds required.

Major uses

More aluminium is produced than any other non-ferrous metal, and none has such a diverse range of uses. The metal has a high thermal and electrical conductivity and a high strength, particularly when alloyed with small quantities of other metals such as silicon and manganese. It is extremely ductile and has found many uses as sheet and foil, and can be cast comparatively easily into all kinds of complex shapes.

Its resistance to corrosion and its attractive appearance makes this metal suitable for many applications in the construction industry. Its lightness (aluminium is less than half as heavy as steel) makes it the most important metal in the aircraft industry. The versatility of aluminium explains its phenomenal growth rate.

It competes with copper in its application as a conductor of electricity, with steel in construction, with zinc for castings and with stainless steel in its corrosion free properties. It is also used for 'de-oxidizing', that is the property of removing oxygen from molten iron and steel. There are few manufacturing and industrial activities which do not have a substantial use for this most useful of metals.

Main market features

Because of the economic and political instability of some of the main bauxite producing countries, many aluminium manufacturers are investigating the possibility of producing the metal from other minerals containing aluminium, of which there are many. These include china clay and shales and good progress is being made in this field. These efforts, however, have been interpreted as a deterrent to any potential threat from major bauxite exporters who might wish to form a cartel to force price increases.

It has become almost fashionable of late for an emerging nation to attach great priority to the building of its own aluminium smelters. This is particularly true of countries in the Middle East, South America and South East Asia. The pattern usually includes the majority shareholding held by the country concerned with a minority interest held by a large aluminium producing company which provides all the technical expertise to construct and run the smelter.

The recent history of the industry has been marked by the failure of the market to predict future consumption trends. This has resulted in a large degree of over-capacity in times of stagnant demand, and shortages in times of high demand. The phenomena is partly due to the long period (about five years) from the date of conception of a new production plant, to the date of commencement of production.

Predictions of the aluminium consumption growth rate vary from 4-8 per cent per annum over the next decade, a very high growth rate for a metal, whichever forecast is used. It should be remembered, that although the price per metric ton of aluminium may seem close to the price of any one of its competitors, the lightness of the metal means that one metric ton will go much further: it has greater volume than any of its competitors, with the exception of magnesium.

Aluminium is essential in the manufacture of highly sophisticated items such as computers and aircraft, but is also used for making commonplace items such as cooking pots. Its consumption, therefore, is not much affected by technical changes in any industry. The market for aluminium is affected by fundamental factors including the supply of raw material, general world economic activity and competition from other metals.

By and large production is expected to keep pace with consumption but a high factor (about 10-15 per cent) of the cost of production is energy. The ever increasing cost of construction of smelters, and of raw materials will inevitably result in an average rise in price. Fluctuations in price from year to year may be more extreme as control of the market shifts from the very few large producers to the many new ones.

Known world reserves (bauxite)

5,842,000,000 metric tons. This is equivalent to 1,168,400,000 metric tons of aluminium metal.

Method of marketing and pricing

The major producers publish an official price for aluminium, the most important of which is the American price. All American producers, at least theoretically, abide by this price. The price is for virgin, unwrought ingots or sows with a minimum purity of 99.5 per cent, CIF major Western ports. There are premiums for a higher purity, different shapes, delivery to works and other ports, or for the delivery of molten metal in specially designed lorries. Most major manufacturers are vertically integrated, producing everything from ingots to foil to manufactured items, often through subsidiaries and associated companies. They are frequently their own largest customers.

Similar official pricing systems exist in the UK, Japan, Continental Europe and other industrialized countries, but in times of glut producers are often forced to sell at discounts on their official price. These discounts are often hidden in extended payment terms or other similar devices.

There is a growing free market in aluminium, and in October 1978 the London Metal Exchange introduced an aluminium futures contract thus adding to the other five metals with which it is concerned. The free market is conducted by international trading houses, which generally have no financial connection with producers. These merchants however, may have a close working relationship with major smelters and could, for instance, conduct 'location swops' or 'time swops' between producers. That is to say they find one smelter with the stock at point A and the sale at point B and another smelter with the stock at point B and the sale at point A. They simply arrange a swop, thus saving transport costs and also collect a commission. If you assume A and B to be times, such as March and July, the swop on this basis is a 'time swop', which saves one company's interest costs. The merchant may also arrange tolling contracts, that is to say he may arrange with the smelter to utilize surplus production capacity to convert the merchant's alumina into ingots for a fixed price per metric ton.

The bulk of free market activity is concerned with the trading of metal from small producers who have no fixed pricing policy and insufficient international marketing arrangements. They must sell to merchants at the best possible price, but in times of shortage they have the advantage in that they are able to obtain higher sales prices than their large competitors who abide by official price policies.

Aluminium metal prices
US cents per pound Producer price

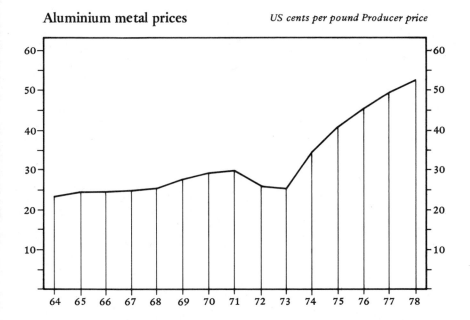

This merchant activity accounts for about 10 per cent of the world's aluminium trade and the prices recorded for these transactions give the best guide to market trends. Aluminium merchants are active in most major commercial centres of the world.

Scrap and secondary aluminium

Many of the useful aluminium alloys can be made from scrap simply by melting it down, adding other metallic alloying agents and perhaps a percentage of pure aluminium, to produce the right specification. This has ensured a very active scrap market where prices seldom dip far below the average. Indeed, the scrap price may be at times as much as 85 per cent of the pure aluminium price.

World secondary aluminium production is approximately one third of primary production.

Antimony

Antimony metal production 1977

metric tons x 1,000

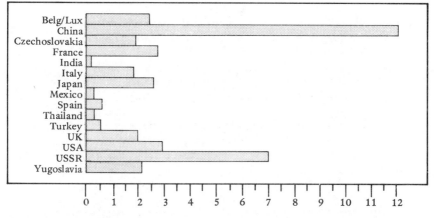

Total world production 40,392 metric tons

Grades available

Antimony metal is normally marketed with the purity between 99 per cent and 99.85 per cent, the most common grades being 99.5 per cent and 99.65 per cent. Particularly attention is paid to the arsenic content which most consumers require to be below .2 per cent. The normal Chinese grade of 99.65 per cent minimum and .15 per cent As maximum has become the most commonly traded grade. The metal is traded in the form of ingots which weigh between 20 and 50 pounds.

Production method

Antimony is produced from its ore, stibnite — antimony sulphide — by roasting to oxide, followed by reduction to the metal by heating with coke in a simple process. Antimony concentrates, derived as a by-product from the production of other metals, mainly lead, are usually in the sulphide form and are dealt with in the same way.

Major uses

By far the largest use for antimony, half its total consumption in fact, is in batteries where it is alloyed with lead. Its effect on lead is to make it more robust and to reduce corrosion by chemicals. The metal is also combined with lead and other metals in the manufacture of solders, print metal, electric cable sheaths, and in certain types of ammunition. Its oxide is particularly important in the manufacture of flame retardent paints and plastics, and as white pigment.

Main market features

China is the largest producer of the metal and has, by far, the largest reserves. Its dominance of the metal market is having a major effect on large Western producers, particularly in the US and the UK. These producers appear to have decided to concentrate on the production of antimony oxide and antimonial lead and to opt out of fierce competition with the Chinese in the metal market. Antimony oxide is made directly from the ore, but, there have been a few temporary occasions in recent times, when market circumstances have allowed producers to manufacture oxide profitably from the metal.

Antimonial lead is an alloy of between 3 per cent and 10 per cent antimony in lead and is used to make the plates in batteries. A large proportion of this material comes from treatment of used and scrap batteries.

Antimony consumption is almost totally dominated by the automobile industry, not only because of its use in car batteries, but also for its use in flame retardent plastics needed in cars.

There are substitutes for antimony in all its major uses. Certain inorganic chemicals are used as flame retardents and are less toxic than antimony. Tin or calcium can and do replace antimony for toughening lead. When considering substitutes, most attention should be paid to the complete substitution of the lead/acid car battery with other types of batteries, which some major manufacturers feel is inevitable.

Although the Chinese are in the position to control the world antimony market, it is assumed that they are keenly aware of potential substitution and will continue to adopt a sensible marketing policy. Prices are expected to be controlled by production considerations rather than by demand, which is unlikely to be affected by anything other than general world economic conditions.

Known world reserves

Approximately 3,550,000 metric tons of which 2,000,000 metric tons are estimated to be in China.

Method of marketing and pricing

There is a very active free market in antimony ore, but price trends for the ore have little direct relation to metal prices. The antimony metal market is rather volatile and attracts speculation, mainly from the merchants who deal in it. This feature is reinforced somewhat by the unpredictability of the Chinese marketing

Antimony prices

£ per metric ton

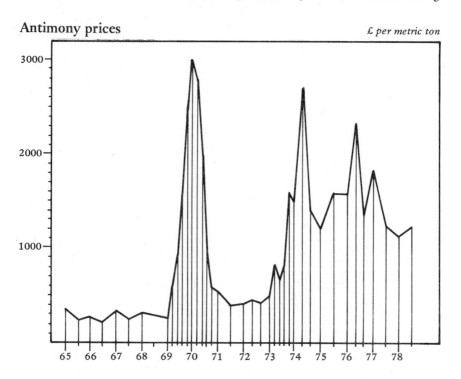

policy. Virtually all Chinese production is sold to international merchants who, in turn, distribute to consumers (many of the major consumers in the free market are in Eastern Europe). Most Western producing countries have high import tariff barriers, but in times of shortage these countries become net importers from China and other producing countries. Periods of shortage are, however, comparatively short-lived. This situation may change as the major Western producers become less interested in the metal market. Prices are recorded of deals between the merchants operating in the world's major financial areas, particularly in Europe. The Western producers fix a price for domestic sales, but these official prices are greatly influenced by free market trends and may frequently be adjusted.

Scrap recovery

Very little scrap antimony is traded, but there is a comparatively large market in antimonial lead.

Arsenic

Arsenic (trioxide) production 1975

metric tons x 1,000

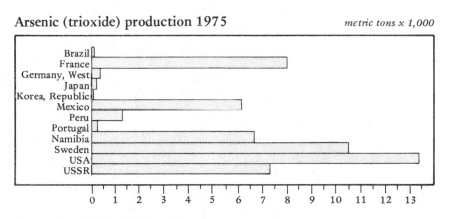

Total world production 54,299 metric tons

Grades available

Elemental arsenic is not strictly a metal in the chemical sense but is referred to as a metal to distinguish it from the oxide in which form it is most commonly distributed. The pure metal is traded in the form of irregularly shaped lumps with a purity of 97 per cent minimum packed in airtight and water-tight drums.

Pure arsenic is rarely traded by merchants but they are a little more active in the arsenic oxide market. The oxide is traded as white powder with a minimum purity of 99 per cent. It is, of course, extremely poisonous and is packed in sealed drums.

Production method

Arsenic is found in many different types of minerals which usually contain other metals particularly lead, copper, zinc, tin, nickel and gold and it is also found together with iron and known as pyrite.

The arsenic is produced from these minerals either as a by-product or co-product of the metal with which it is associated.

In most smelting methods the arsenic is collected as a flue dust in the metal

70

refining process. It is then added to pyrite or galena and heated until the arsenic comes off as a gas which is then collected.

Major uses

About 97 per cent of arsenic production is in the form of oxide which is mainly used to make a large variety of pesticides. Other uses are as a decolourizer in the glass industry and in metal refining.

Pure arsenic is used as an addition to lead in alloys to increase hardness and to copper where it increases corrosion resistance. Copper arsenic pipes were used extensively in steam train locomotive fire-boxes but this use has almost disappeared with the passing of the steam train era.

Main market features

Arsenic production is widely spread over the globe, and the oxide is rather inexpensive. This situation obviously does not lead to a very exciting market. Sweden, one of the world's largest producers, distributes its material throughout the world on a flexible producer price basis which has traditionally set the price trend for most other trades. There has been a huge drop in both production and consumption of arsenic over the last decade due to environmental considerations,

Arsenic metal prices

£ per metric ton. Swedish price delivered UK

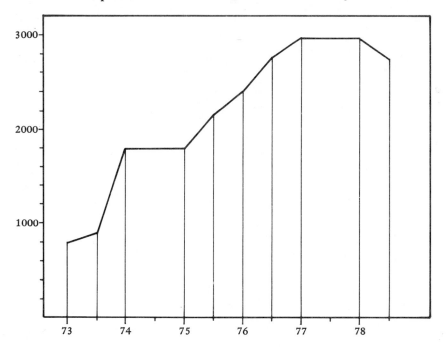

which have further discouraged free market activity. As world trade contracts, producers are more inclined to discount their own official producer price.

Pure arsenic is much more expensive than the arsenic oxide, but most oxide producers are not interested in producing the elemental form from the oxide. This has led to short periods of active free market trading in times of shortage but demand is somewhat erratic and no sustained period of shortage is possible because so little of the oxide is used to produce the metal.

Method of marketing and pricing

Pure arsenic metal prices are closely controlled by producers but a free market does emerge at premium prices in short periods of high demand. Merchants are only occasionally involved in the oxide market but trade is now highly competitive between producers.

Beryllium

Beryllium ore production (metal content) 1977 *short tons*

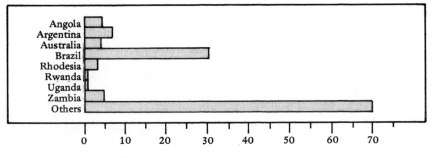

Total world production 125 short tons

Grades available

Pure beryllium metal is rarely traded. Apart from being an extremely toxic metal, which makes handling very dangerous, it is only used commercially in the form of an alloy with other metals. It is therefore safer and easier to market beryllium in the form of a master alloy (an alloy with a high beryllium content compared with its content in the final form) which can then be diluted in the melt. The most common of these master alloys is beryllium-copper with a content of up to 5 per cent beryllium.

Production methods

There are several methods of extracting beryllium oxide from beryl, its ore. They are all comparatively simple chemical methods.

Beryllium fluoride is used as the starting material for producing the metal. This is done by reducing the fluoride in an electric furnace.

Major uses

Most beryllium is used in the form of beryllium-copper, a very important copper alloy used to make marine propellors, springs, aircraft engine components and electrical contacts. It is also used to make non-spark producing tools for use in coal mines and oil refineries, where there is a danger due to inflammable gases. There are minor applications for other beryllium alloys such as beryllium-nickel, and some pure beryllium is used, mainly in nuclear energy plants.

Main market features

Although beryllium ore is produced in many countries, the production of the metal is in very few hands and these are mainly in the US. These producers effectively control the market. This situation is understandable since there are dangers involved in producing the metal due to toxicity and, in addition, the market is small. Producers are rarely asked to supply pure beryllium metal, but conduct most of their business in master alloys.

There is almost no merchant activity in this metal.

Method of marketing and pricing

The main producers have been successful in controlling prices and there is practically no free market outside this system.

Beryllium ore prices

US cents per metric ton unit

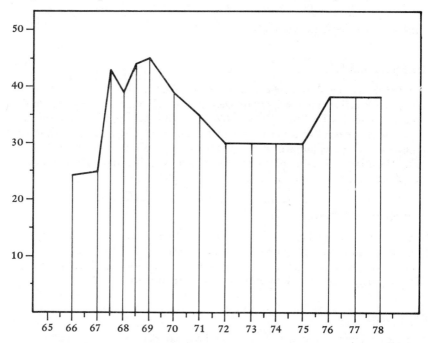

Bismuth

Bismuth (contained in ore) production 1976   *metric tons x 100*

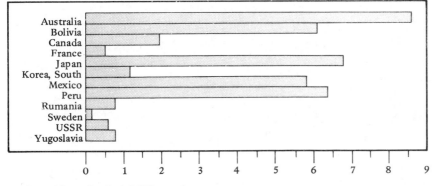

Total world production 3,973 metric tons

Grades available

Bismuth is marketed in the form of ingots usually weighing 10 kilos each and packed in wooden boxes each weighing 100 kilos, or in the form of needles or granules weighing between 3 and 5 grams each and packed in 100 kilo steel drums.

Purity is 99.95 per cent minimum up to 99.99 per cent minimum. Some small producers market bismuth in a low refined state (around 96 per cent purity) known as bismuth bullion, usually contaminated with lead. Consumers of the metal who produce pharmaceutical products may demand a low tolerance of trace elements.

Production method

Usually as a by-product of lead or zinc manufacture but also as a by-product of copper, silver, gold and tin refining. The bismuth rich concentrates, usually sulphides, are oxidized in a simple and inexpensive process to the metal. It is then upgraded by skimming and recrystallization and treatment with flucculents to produce metal with a purity of 99.99 per cent suitable for most industrial and pharmaceutical uses.

Major uses

a) *Pharmaceutical* Some salts of bismuth have been found beneficial in the treatment of indigestion and other minor ailments of the alimentary canal and these account for over half the world's consumption. The French, however, who have traditionally consumed almost one third of the world's production for this purpose, decided in 1974 to make bismuth products available to patients only on prescription which has had the effect of reducing that country's consumption by over half. Applications have been found for other bismuth salts in cosmetics.

b) *Fusible Alloys* Bismuth is used as an alloy with lead, cadmium and other metals in the manufacture of fusible alloys, the bulk of which is used as jigs for making parts in the aircraft industry, and also in the manufacture of optical lenses. Other uses are found for these alloys in press moulding machines, the dyeing of textiles, fire sprinkler systems and solders.

c) *Catalysts* Bismuth salts are used in a range of catalysts, particularly in the manufacture of acrylonitrile, a raw material for a man-made fibre.

d) *Alloying agent* Bismuth is added to certain aluminium alloys to improve machinability and also to other metals for specialized uses.

Consumption by use (approximate)

Fusible Alloys	22 per cent
Metallurgical additives	18 per cent
Pharmaceutical	50 per cent
Catalysts	6 per cent
Other Uses	4 per cent

Main market features

Consumption of bismuth has been subject to large fluctuations. The popularity of bismuth as a catalyst in the 1960s and the legislation concerning medical products in France, together with the fortunes and misfortunes of the world's aircraft industry, have all contributed to make the bismuth market one of the world's most volatile. The price volatility itself has contributed to nervousness amongst potential consumers, but bismuth has some unique physical characteristics which must find a continuing interest in industry, especially at its current very low price.

There is no official market for bismuth. The major producers fix an official price, but in times of shortage the material is rationed to consumers and the official price is often raised. In times of glut the producers offer discounts on the official price.

It is true to say that since the late 1960s the major producers have lost control of the pricing of this metal and free market forces have taken over. Although perhaps only 25 per cent of the world's production is distributed by merchants, it is their transactions that give the best guide to the price movements.

The merchants who specialize in bismuth, trading mainly in New York, London, Brussels and Dusseldorf, trade the metal throughout the working day, buying surplus stocks from producers and distributing to consumers. These merchants often speculate by holding stocks or 'going short' of the metal and therefore often trade between themselves. They trade in minimum one metric ton lots, but average transactions would be five metric tons.

There is very little scrap trade in bismuth, except for some adulterated alloys that may need extensive refining.

Estimated world reserves

93,000 metric tons

Bismuth metal prices   *US dollars per pound*

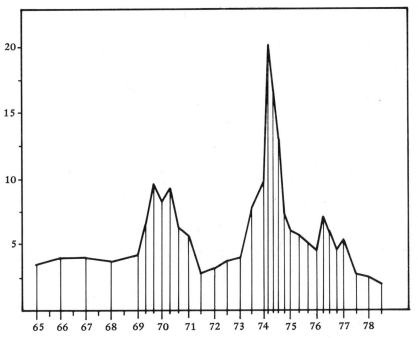

Cadmium

Cadmium production 1977

metric tons x 100

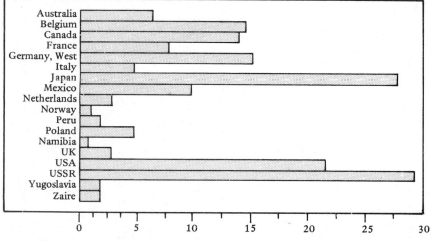

Total world production 16,630 metric tons

Cadmium consumption 1975

metric tons x 100

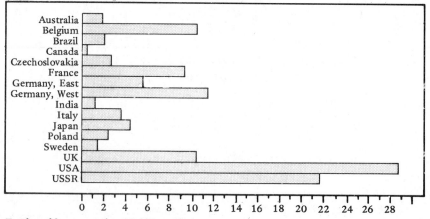

Total world consumption 11,638 metric tons

Grades available

Cadmium is marketed in forms determined by its use; balls approximately 2 ins in diameter for plating, sticks about 10 ins long and ½ in diameter for use in chemical works and ingots which are used in alloys. Purity is most commonly 99.95 per cent minimum, but often 99.99 per cent minimum which is required for some chemical purposes.

Production method

Almost all cadmium is produced as a by-product of zinc production, but it may also occur in recoverable quantities in lead and copper ores.

Cadmium rich residues are collected in base metal refining processes usually in the form of flue dusts. It is then dissolved in sulphuric acid and the metal is either separated electrolitically or by chemical methods. Further purification can then be carried out by distillation.

Major uses

Cadmium has four major uses − in plating, pigments, polyvinyl chloride stabilizers and in alloys. Plating is the most established use. Cadmium plating of steel products is carried out to ensure a firmer base for another plated metal, like chrome, which would be electroplated over the cadmium to counteract certain corrosive elements, such as sea water. Mechanical and electronic equipment used in coastal or marine work often requires cadmium plating.

Cadmium salts which are made directly from the metal are used to make pigments for ceramics and plastics and are used as stabilizers in the manufacture of PVC. Cadmium is highly toxic and this prevents it being used in any material that comes into contact with foodstuffs.

The metal is also used in combination with tin, lead and other metals to make a variety of solders, print metal and fusible alloys used in sprinkler systems and infusible jigs.

Main market features

Cadmium production is as widely spread as zinc production and no single producer is in a position to control the market. Like many similar metals the North American producers publish a producers' price for home buyers and for exports. Most other producers, however, don't seem to set a high priority on controlling cadmium prices, and sell at the best price they can to international merchants who specialize in the trading of minor metals.

Known world reserves

Approximately 550,000 metric tons

Method of marketing and pricing

Alternative metals are available for all cadmium's major uses. This fact, combined with the danger of toxicity sets a limit to its competitiveness and price. This has not prevented periods of scarcity of the metal in recent times with consequent high prices. As a by-product metal, it has an upper limit of production capacity and any sustained shortage would encourage users to switch to alternative materials. It often takes some time, however, for a consumer to change production methods and in the past, scarcity was too short-lived to encourage wholesale substitution. This threat, however, will continue to exist.

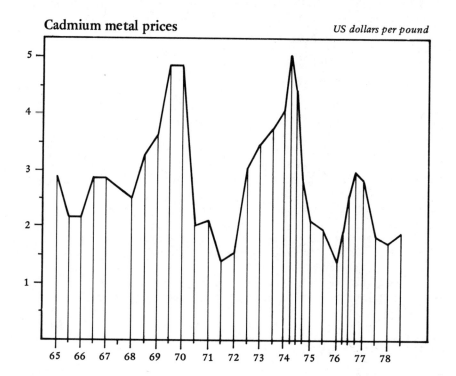

Cadmium metal prices *US dollars per pound*

Cerium

Cerium (monazite ore) production 1975　　　　　*short tons x 1,000*

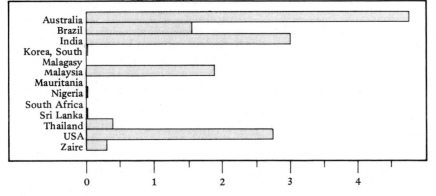

Total world production 11,895 short tons

Grades available

Pure cerium metal is of no interest to the metal dealer and is only produced for laboratory uses and research purposes.

Mischmetal, however, a mixture of rare earth metals containing about 50 per cent cerium, is regularly traded in the form of small ingots, pellets and notched bars.

Production methods

There are several ores containing cerium and other rare earth elements, the most common of which is monazite. It is also found in commercially recoverable quantities in xenatime, bastnaesite and loparite. Monazite is found in a beach sand ore and the rare earths are recovered here as a by-product of titanium production. Rare earth ores are separated from the beach sand by magnetic and electrostatic means. They are concentrated by flotation and the metal (mischmetal) is produced by electrolysis of the rare earth chlorides.

Major uses

Cerium compounds, particularly the oxide, have several important uses as filters and de-colourizers in the glass industry and as catalysts in petrochemical production.

Mischmetal alloyed with iron is used as lighter flints, but its main metallurgical use is as an addition together with magnesium to certain cast irons known as nodularized irons. It is also added to certain types of steels to increase strength and ductility.

Main market features

There are very few companies engaged in the manufacture of cerium products and mischmetal. These companies are mainly in industrialized countries and have enough control of the market to maintain an effective producer price in times of normal demand. There have been a few brief periods in the last two decades when unexpectedly high demand has led to an active free market in mischmetal. During these periods major producers have rationed material to their regular customers. Consumers, especially in the cast iron industry, have then appeared as buyers often at quite substantial premiums to the producer price.

Temporary shortages of mischmetal production have also resulted from inadequate supplies of suitable ore or intermediate raw materials.

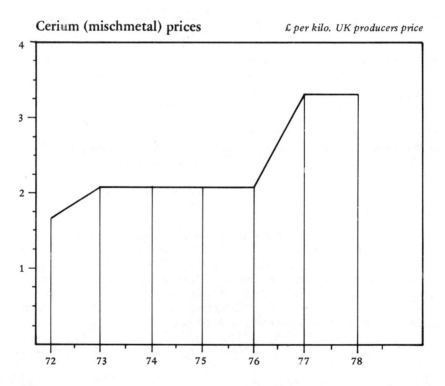

Cerium (mischmetal) prices

£ per kilo. UK producers price

The effect of small mischmetal additions to iron and steel is a rather complicated and interesting branch of metallurgy and new uses for this compound are expected to increase demand in this field.

Known world reserves

Cerium containing ores are 10 million tons of rare earth oxide contained, about half of which is estimated to be in the US.

Method of marketing and pricing

As has been mentioned, a few producers of cerium products operate in industrialized countries where they sell direct to consumers, especially on their domestic market, at fixed producer prices. During the comparatively short-lived periods of shortage, however, mischmetal has attracted attention from international merchants operating at premium levels.

Chromium

Chrome ore production 1976

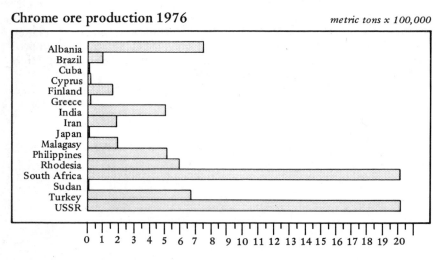

Total world production 8,406,500 metric tons

Grades available

Only a small part of chrome produced is converted into chromium metal. The bulk of chromium is used in the form of its ore (chromite) or ferro-chrome. Chromite ore is simply mined, cleaned, crushed and graded. Ferro-chrome, the alloy of chrome and iron is produced in several grades. Charge-chrome (containing about 55 per cent chrome), high carbon ferro-chrome (5-8 per cent carbon); and low carbon (0.1 per cent carbon maximum). Of these three, the cheapest is charge-chrome. Another inexpensive chrome containing alloy is ferro-silicon chrome.

Chromite is invariably sold in bulk but ferro-chrome is sold in a lumpy form and may be transported in bulk or packed in steel drums.

Production methods

Chromite is melted in an arc-furnace together with iron to produce ferro-chrome. Chromium metal is produced either by electrolysis of a chromium solution or by alimino thermic reduction, whereby chromium oxide is mixed with aluminium powder and ignited, thus reducing the oxide to chromium metal.

Main market features

Perhaps the most outstanding feature of the chromium market is the fact that the US, the world's biggest end-user of chrome, has no indigenous production.

Chromium is one of the world's most important industrial metals and recent events have once more underlined the potential vulnerability of the US chromium consumer to foreign political events.

South Africa and the USSR produce between them about half the world's chromium in the form of chromite ore and ferro-chrome. Together they represent the bulk of the world's exports.

It can be understood that the US is not too happy about being dependent on either the USSR or South Africa for the supply of such an essential raw material.

To add to US difficulties Rhodesia is a significant source of chrome (approximately 8 per cent of the world supply). President Carter in his early days of office repealed the 'Byrd Amendment' which allowed Rhodesian chrome to be imported into the US in spite of UN resolutions. The repeal of this amendment not only means that Rhodesian chrome cannot be imported into the US but also that any steel containing Rhodesian chrome cannot be imported either. It should be noted that this move coincided with a huge over-supply of chrome in the world market but that it has further eroded the US position should supplies once more become tight.

Another important feature of the chrome market is the technical innovation in the production of stainless steel. The comparatively new AOD (Argon-Oxygen Decarburising Vessel) has been found to be the most efficient method of producing stainless steel and among its technical advantages is its ability to utilize lower quality raw materials including chrome.

In the past, ferro-chrome producers were encouraged to reduce the carbon content in their product, an extremely expensive operation. With the advent of the AOD, carbon content can be substantially reduced in the steel making process. Ferro-chrome producers, aware of this change, have swung over on a large scale to the production of charge chrome with a chrome content of about 55 per cent as opposed to the 70 per cent chrome content in standard, medium, and low carbon ferro-chrome. The carbon content of charge chrome can be as high as 6-8 per cent without worrying stainless steel producers using AODs.

Major uses

About 20 per cent of the chromite mined is used directly in refractories, foundry mould sand and other non-metallurgical uses. Its other two major uses are in alloy form in steels and for plating. Chrome is added, in the form of ferro-chrome, to steel to produce alloys including stainless steel. There are a large range of very useful chrome steels but by far the most common of these is normal stainless steel containing both chrome and nickel.

Pure chromium metal is used to manufacture many types of non-ferrous alloys and super alloys, used in magnets amongst other things. Chromium is plated on to such items as automobile trim and cutlery from a bath of chromic acid which is made directly from chromite.

Known world reserves

Approximately 775 million metric tons of chromite ore. 575 million metric tons are estimated to be in South Africa.

Method of marketing and pricing

World trade in chromium can be confused by the complications of tariff barriers and quota systems in individual markets. However, the price trend is set by South African producers and followed competitively by Russian exports. Most business is conducted by producers and their agents but there is an extremely active free market conducted by large merchants utilizing particularly Russian supplies and material from the smaller producing countries. Business is highly competitive mainly because the important trade is conducted with very large steel mills which usually place business on a long-term basis. There are many nationally based merchants who trade on a smaller scale with small specialist steel works and foundries.

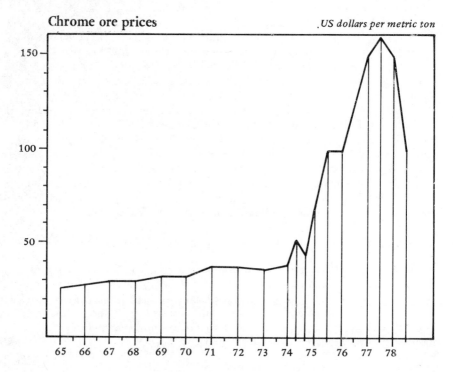

Chrome ore prices   *US dollars per metric ton*

Cobalt

Cobalt metal production 1976

metric tons x 1,000

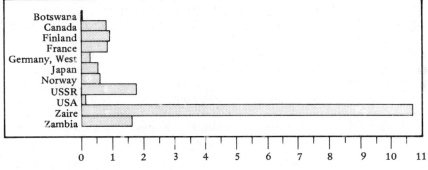

Total world production 18,360 metric tons

Grades available

The most commonly traded form of cobalt is broken cathodes, small flat pieces of the metal produced by crushing a cathode of the brittle pure metal. Granules are a less common form of the pure metal but both are normally distributed with a purity of 99.6 per cent minimum.

Production method

Cobalt is almost universally produced as a by-product, or co-product, of either nickel or copper, except in Morocco where the metal is mined on its own. The cobalt rich concentrates are normally collected during the electrolysis of the base metal concerned and is itself produced by electrolysis.

Major uses

The two major uses for this metal are in what is termed super alloys used extensively in the aircraft industry (it is impossible at present to make a modern jet engine without the use of cobalt alloys) and in turbines. Super alloys are usually high alloy steels but may contain only other non-ferrous metals. The

87

second major use is again in alloys, consisting of cobalt and other non-ferrous metals used to make magnets and other electromagnetic components in the electronics industry.

Cobalt alloys are extremely useful in applications where strength at high temperatures is required. There are certain important applications for cobalt salts, particularly the oxide, which is used as a catalyst in the petrochemical industry and as pigment in ceramics and enamels.

There is a very minor use for cobalt isotopes in the treatment of cancer. Pure cobalt powder has an application in sintered products such as machine tools and ammunition when combined with other refractory metals.

Main market features

Many of cobalt's applications are in highly strategic areas, such as its use in aircraft engines and fuselages. It is interesting to note that although the US consumes over one third of the world's production, directly or indirectly, it has almost no production of its own.

About 65 per cent of the world's production comes from Zaire as a by-product of copper. Zairean production has been disrupted twice in recent history by civil war. Until these disruptions occurred, however, the cobalt market had been extremely stable due particularly to the combined efforts of the Zairean mining authorities and the Belgian refiners who control the pricing

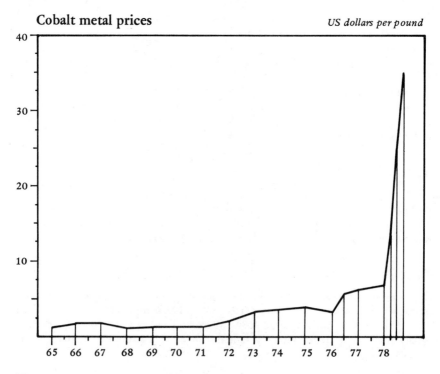

Cobalt metal prices   *US dollars per pound*

and distribution of the Zairean metal. Disruption of supplies has demonstrated the vulnerability of the cobalt market and caused massive fluctuations in the free market price.

The world's second largest producer Zambia is also operating in a very politically unstable area. There has been speculation that the American stockpile may be used to alleviate temporary shortages but this is unlikely as its stock is already considered to be at a low level. All these factors are likely to ensure important shortages of supply, efforts to substitute cobalt with other metals, a search for new commercially viable sources and a volatile market in the foreseeable future.

Known world reserves

Approximately 1,100,000 tons of which 700,000 tons are estimated to be in Zaire.

Method of marketing and pricing

Until the disruption of Zairean production, 90 per cent of all cobalt produced was marketed on a producer price method. Accelerating prices have encouraged a very active free market which will certainly grow in significance. Traders obtain supplies from small producers and consumers with surplus stocks. Supplies from these sources, however, are likely to be more difficult to obtain in the future but rising prices may lead to production from small mines or from ores with low cobalt contents, thus in part alleviating future shortages.

Columbium or Niobium

Columbium ore production 1976

metric tons x 1,000

Country
Australia
Brazil
Canada
Malaya
Mozambique
Nigeria
Portugal
Rwanda
Thailand
Zaire

0 1 2 3 4 5 6 7 8 9 10 11 12 13 14 15 16 17 18

Total world production 22,000 metric tons

Grades available

Pure columbium metal is a rather rare item accounting only for a tiny fraction of total columbium consumption. Its most commonly traded form is ferro-columbium (its alloy with iron) containing from 60-70 per cent columbium. In this form it is sold as irregular lumps packed in drums.

Production method

Columbium is extracted from its naturally occurring ore pyrochlore and from the ore columbite which strangely contains both columbium pentoxide and tantalum pentoxide. If the tantalum content exceeds the columbium content the same basic mineral is known as tantalite. The ore is concentrated by a combination of wet gravity, magnetic and electrostatic processes.

In columbite and tantalite, columbium oxide is separated from the tantalum oxide first by conversion to the fluorides then by a solvent extraction method. The columbium is then reconverted to the oxide by calcining.

Ferro-columbium is produced by the thermic method, that is burning the pentoxide with aluminium powder in the presence of iron powder. Columbium metal can be produced by reducing columbium chloride with magnesium metal.

90

Major uses

Almost all columbium is used in the form of ferro-columbium and more rarely in the form of the pentoxide in the manufacture of alloy steels. These are in turn used for structural purposes in buildings, bridges etc, for heavy mining equipment such as rock cutters and for machine components where shock resistance is required. A recent growth area for columbium consumption is in the manufacture of gas and oil pipeline steels. Minor uses for the metal occur in the nuclear energy and electronics industries.

Main market features

Brazil is responsible for producing about 75 per cent of the world's supply of this metal and holds a similar proportion of the world's known reserves. Columbium producers have so far had a very responsible attitude to pricing, with prices rising steadily in line with the growth of consumption and inflation. The effects of the 1974 oil price rise has, however, greatly encouraged the construction of oil pipelines producing an unforeseen demand for columbium. Other metals used for pipeline steels, vanadium and molybdenum have shown marked price rises for the same reason. Brazil is in a uniquely powerful position to raise prices if demand ever justifies such a move and there is little competition from other

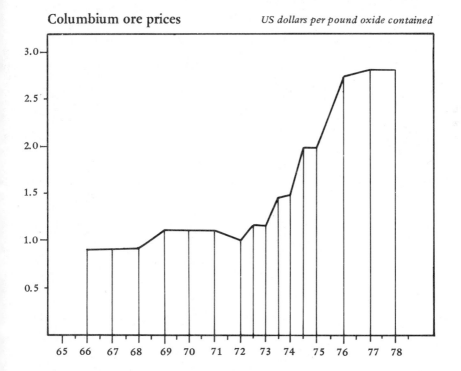

Columbium ore prices   *US dollars per pound oxide contained*

columbium producers. But the competition from vanadium, which can be substituted for columbium in many alloys, needs to be remembered.

Known world reserves

5,100,000 metric tons of columbium (metal content) of which approximately 4 million metric tons are estimated to be in Brazil.

Method of marketing and pricing

Columbium has not been a particularly popular item with metal merchants in the past due mainly to the efficient and flexible pricing policy of main producers who are able to adjust prices and stocks according to demand. They have none of the problems of molybdenum producers, for instance, whose metal is largely a by-product material. Demand in recent times has been stronger, however, and the efforts of the producers to strike a compromise between maintaining a steady market and meeting extra demand, may stimulate a small premium market.

Most producers either sell directly to consumers or have local agents to market their product. International merchant activity is only apparent during temporary shortages of material.

Copper

Grades available

Pure copper is most commonly traded in two forms, cathodes and wirebars.

A cathode is the pure copper slab which results from the electrolytic refining process. Its usual dimension is about 36ins x 36ins x ½in. It has a bubbly surface and a minimum purity of 99.9 per cent copper. This purity includes a small silver content.

Wirebars which are generally cast from cathodes are the elongated ingots used as a starting rod for drawing copper wire. Its standard weight is between 90-125 kg, again with a purity of 99.9 per cent copper plus silver.

Many producers offer grades with certain chemical characteristics such as HCFR (High Conductivity Fire Refined) which is used for applications where high electrical or thermal conductivity is required. Producers also offer fabricated forms such as wires, tubes, sheets etc, but these markets are more the concern of the stockist than the trader or broker.

Production method

There are many types of copper ore but commercially recoverable deposits are either sulphides or less commonly, oxides. Occasionally, copper is extracted from complex minerals containing other metals such as lead or zinc.

The ores are concentrated by the normal methods of crushing and flotation. Copper salts may be extracted by leaching, that is to say treating the ore with an acid that will preferentially combine with the copper and the resulting copper rich solution can, in turn, be treated to extract the metal. Leaching is particularly useful for refining low grade ores or mine waste. Many copper ores contain other useful non-ferrous metals such as molybdenum, cobalt and selenium and methods to extract these metals in refinable form are incorporated in the copper refining process.

The ores may first be roasted, if the required desulphurization is impossible in the smelting process. The smelter produces an impure form of metal known as blister copper which is cast into large flat ingots. These are used as anodes for the electrolytic refining process which is carried out in the normal way using thin sheets of pure copper as cathodes, onto which the copper is plated.

Copper mine production 1977

metric tons x 1,000

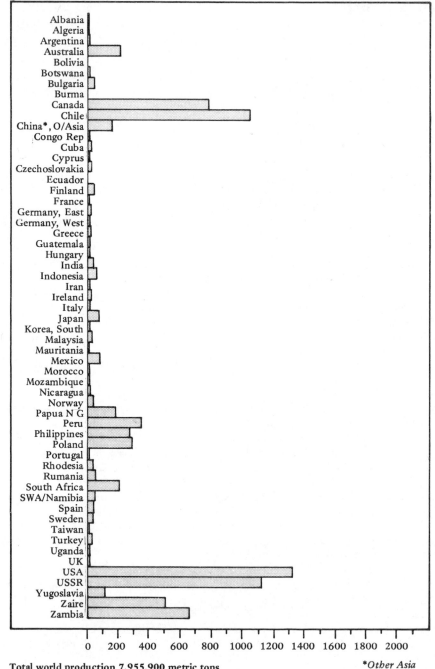

Total world production 7,955,900 metric tons

*Other Asia

Copper refined consumption 1977

metric tons x 1,000

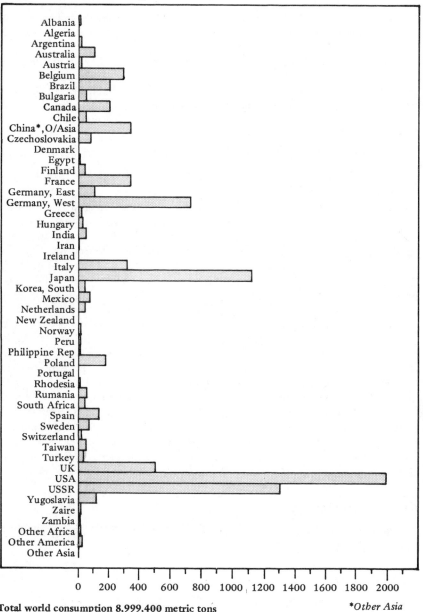

Total world consumption 8,999,400 metric tons

*Other Asia

Major uses

Copper's most familiar uses are as wire for conducting electricity and as its alloy brass for screws and fittings. It is in the electrical industry where most copper is consumed.

Other important uses are as pipes for domestic water supply, in the chemical industry as heat exchangers which include car radiators, refrigerators and industrial cooling systems. The above uses exploit the excellent electrical and thermal conductivity of copper, which are important elements in almost all industries although automobile and electrical cable manufacturing are particularly important consumers.

Another major use is in the manufacture of coinage.

Main market features

The movement of copper prices provides a good indicator of the world's economic state, since it is so widely used throughout industry and elsewhere.

The marketing of copper is greatly influenced by the fact that an increasingly large proportion of supplies, and future reserves, are concentrated in the developing countries, particularly in the African copperbelt running from Zambia to Zaire, and in South America from Chile to Peru. New deposits are likely to be found in the remoter areas of the world, since those in developed areas have been exploited during the long period in which copper has been a major metal utilized by mankind and these are gradually becoming exhausted. For the moment the world's two largest producers are the United States and the Soviet Union, but they are also the largest consumers and are, therefore, generally net importers. Canada is another major producer and exporter and Australian production is increasing. Recovery of scrap copper is growing in industrialized countries, providing an additional and major source of supply with a totally different cost structure. A new source of future supply is to come from nodules on the seabed, which have been found to contain an important proportion of copper. But the immediate pattern is for the exploitation of lower grade ores and deposits in remoter areas, as well as increasing supplies of scrap copper.

Known world reserves

275 million metric tons excluding potential re-usable scrap which amounts to some 20 per cent of known reserves.

Methods of marketing and pricing

Copper is sold mainly by annual supply contracts, negotiated between consumers and producers. An increasing quantity of copper exports produced in developing countries has come under the control of state organizations selling on a national basis. Developing countries have been seeking to expand their domestic smelting and refining facilities, but copper ore concentrates are sold to smelters in the developed world as well as blister copper. In refined form copper is sold either as wirebars or as cathodes. Buyers of primary and scrap copper are basically fabricators, who turn it into semi-manufactured forms for sale to industry. Sales often pass through several hands first, including government organizations and merchants.

In the US, the copper industry is to a great extent vertically integrated with the mine producers, who also own smelting and fabricating facilities. This has enabled the US to retain the domestic producer price system, under which the producer fixes the price at which he is prepared to supply copper under long-term contracts. But in 1978 the biggest US producer, Kennecott, decided to abandon its producer price and instead base prices on the 'free' market, New York Copper Exchange (Comex). It was followed by Anaconda, and other US producers have been forced to introduce much more flexible pricing methods.

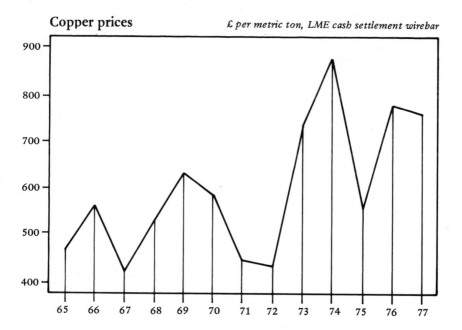

Copper prices *£ per metric ton, LME cash settlement wirebar*

Gallium

Gallium production 1975

kilos x 1,000

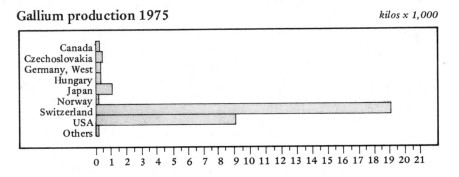

Total world production 30,700 kilos

Grades available

Gallium of a very high purity between 99.9 per cent and 99.9999 per cent is required for most uses. It is usually sold in the form of small ingots.

Production method

All gallium is produced as a by-product of aluminium and zinc production.

It is only extracted from a small proportion of aluminium or zinc ore due to the expense of the process and the low demand in terms of tonnage for gallium. It is extracted from the alumina by precipitation from a caustic solution, then by electrolysis with gallium deposited at the cathode.

Gallium is extracted from zinc concentrates by a simple chemical process and electrolysis, using platinum electrodes.

Major uses

Gallium has two main uses; in electrical semi-conductors and in measuring devices. Its use in semi-conductors represents over 90 per cent of consumption.

Main market features

The fact that gallium is a by-product does not mean that the potential for production is limited by the production of the base metal ore from which it is produced, since so little of the main ore is made available for gallium extraction.

Any increase in demand could be met, but time might elapse before plant could be prepared to do this work.

The semi-conductor industry is extremely fickle and regularly changes its requirements for raw material. This means that alternatives for gallium may be found after some technical innovation. On the other hand, it is quite likely that new uses for this metal could arise in some sophisticated application.

No single producer has the power to exercise any great influence over the market, but as the market is so small and technical, prices remain high to compensate for the stringent level of quality control, and of technical advice that must be offered by any producer.

Method of marketing and pricing

There are very few producers of gallium and all are in industrialized countries where the bulk of consumption occurs. Most producers sell at similar prices direct to consumers. The majority are also large producers of aluminium or zinc and have good marketing arrangements in consuming countries.

Prices vary according to grade, but consumers are as interested in technical reliability as much as price, so there is little scope for merchant activity, except in brief periods of shortage or of selling to state buying corporations.

Germanium

Germanium ore production 1977
pounds x 10,000

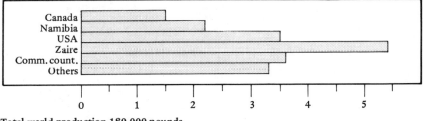

Total world production 180,000 pounds

Grades available

Almost all germanium is sold in the form of very high purity ingots or in single crystals. Purity is usually not less than 99.999 per cent minimum, but grades are determined normally by electrical resistivity, the most common being 30 OHM CMS or 50 OHM CMS.

Production method

All germanium is produced as a by-product of other base metal production, usually zinc or copper.

In zinc ore, germanium oxide is volatilized during the refining process and collected electro-statically.

In copper refining the germanium rich residue is collected magnetically as flue dust and the germanium salt is leached out chemically, then smelted to the metal in a hydrogen atmosphere.

A high degree of purity is required for almost all uses of germanium. This is done by standard zone refining techniques in an electric induction ring.

Major uses

Over 90 per cent of germanium consumption is in the electronics industry for various forms of semi-conductors, such as diodes, transistors and rectifiers.

Other small quantities are used in the manufacture of specialized glass and for certain solders.

Main market features

Although germanium occurs in incredibly minute proportions in the ores from which it is recovered, there would be little difficulty in recovering it from other sources such as power station flue dust where it occurs in even smaller concentrations but in greater quantities. This could only happen if for some reason demand were to grow or if present production were to be disrupted.

Much higher prices would be necessary to make this alternative supply commercially viable. The electronics industry seems regularly to go through a technical revolution and with the advent of new semi-conductor materials, there may be a much lower demand for germanium.

Production costs are really the only factor to consider when accounting for its price, as production can be increased to meet any foreseeable demand. Short periods of increased demand have led in the past and may in the future lead to temporary but sharp price increases.

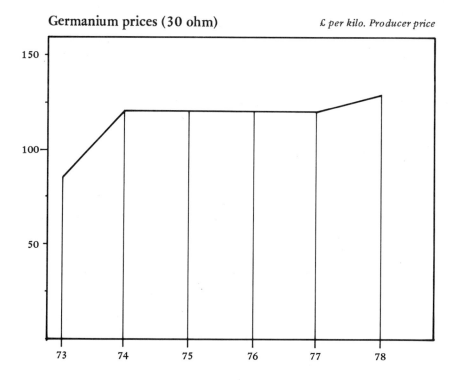

Germanium prices (30 ohm) *£ per kilo. Producer price*

Known world reserves

These are approximately 1,500 tons but this estimate is based on extrapolated production statistics rather than geological data.

Method of marketing and pricing

Because the germanium market is so very small and fickle and marketing so technically orientated, producers need to be well rewarded for their efforts. Prices therefore put germanium into the precious metal class.

Prices vary according to grade and most producers, with perhaps the USSR as the exception, sell at much the same levels. USSR sales are intermittent and as the price of the metal represents only a small proportion of the value of the component in which it is used, most consumers prefer to rely upon producers who have the best continuity of supply and sound technical reliability.

Merchants are rarely involved in the trade of germanium, except between state-controlled producers and consumers.

Indium

Indium production 1974 *troy ounces x 10,000*

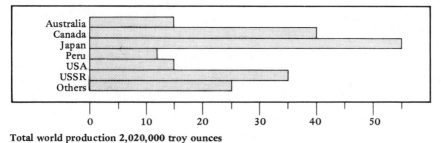

Total world production 2,020,000 troy ounces

Grades available

Indium is normally traded in the form of bars or ingots with a minimum purity of 99.99 per cent but some producers still offer a 99.97 per cent grade which is traded at a discount to the normal price.

Production method

All indium is produced commercially as a by-product of zinc and lead production, but only some zinc and lead ores contain indium in recoverable quantities.

The indium could be recovered by a number of methods, the most common being distillation but it is also recoverable by leaching and electrolysis.

Major uses

Indium has a wide range of uses: as a thin alloy layer in high performance bearings, mainly for sports cars; in the electronics industry, as a component of transistors and in computer equipment; in low melting point alloys and in solders.

Some of these low melting point alloys are used for glass lens grinding and polishing, and as plugs for fire sprinkler systems.

Indium is also used in catalysts and in the purification system for the

103

manufacture of certain organic chemical products, particularly man-made fibres. The metal also has many minor uses in research and in laboratory techniques.

Main market features

Practically all the uses for indium are either technologically sophisticated or obscure. The electronics industry is notorious for rapidly revolutionizing systems and completely changing its requirements for raw materials. Indium has suffered setbacks in demand as result of this over the last few years.

Demand from other areas has fluctuated also but because there are so many uses for this metal, the net effect has been to keep the market surprisingly steady.

The mining of zinc and lead ores, which contain useful and expensive by-products is an obvious way for mines to increase profitability. But such ores are comparatively rare, and over the years the content of these metals has diminished, just as the cost of extracting them has increased. Because indium is used in the more technical areas of industry, it is believed that more new uses for the metal will be found as industry itself becomes more technical. This should result in increasing prices over the long-term with perhaps even larger fluctuations of price in the short-term.

Indium metal prices

£ per troy ounce

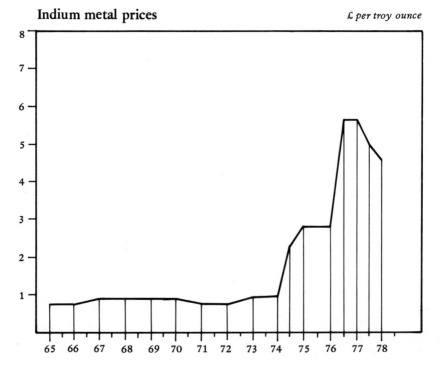

Method of marketing and pricing

Canada is the largest producer of indium and together with the producers in the US, maintains a strong but responsible influence on the market. As with many other metals, there are very few large consumers of indium and a multitude of small consumers. The large consumers almost invariably sign their supply contracts with one or other of the large North American producers. This is understandable, as a continuity of supply as well as technical reliability are almost as important as the price.

The price paid by these consumers provides a pricing basis for other producers when selling to smaller consumers but a discounting or premium system exists depending on the state of the market.

There is a small but fairly active free market conducted by the few merchants who specialize in this type of rare metal. Their main source of material is the USSR which conducts a sporadic sales policy. Supplies are also available from Japanese and South American producers who prefer to sell to international merchants rather than to set up their own marketing operation which would be disproportionately expensive owing to the large number of smaller consumers who need to be serviced.

Lead

Lead mine production 1977

metric tons x 1,000

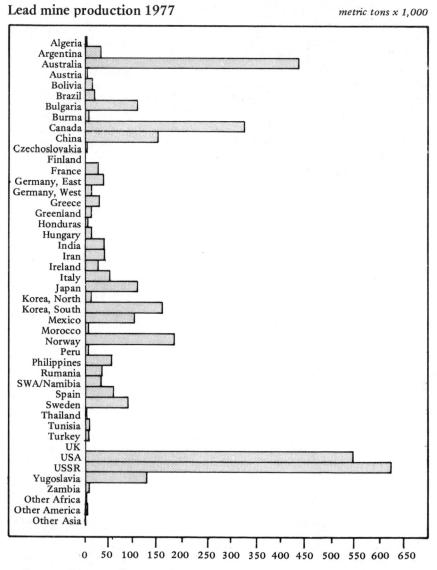

Total world production 3,662,000 metric tons

Lead refined consumption 1977

 *metric tons x 1,000*

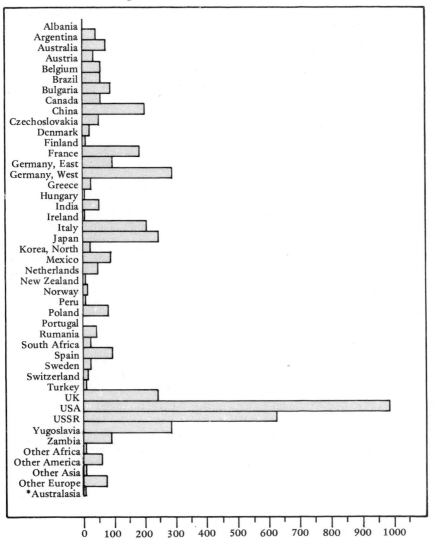

Total world consumption 4,419,300 metric tons   **Other Australasia*

Grades available

The most commonly traded grade of refined pig lead is 99.97 per cent minimum. This is the standard minimum grade which can be traded on the London Metal Exchange. There is a large trade in secondary lead (remelted from scrap) which normally has a purity of 99 per cent minimum.

Production method

Lead is found in several minerals, but its most common ore is galena (lead sulphide). Commercially viable lead ores can also be associated with certain zinc bearing minerals. Preparation for smelting varies with the grade of ore. Some high grade material may not need treatment, but low grade ores are concentrated by flotation. The high grade ore or concentrate is roasted in air to remove the sulphur, then smelted in a blast furnace or open hearth furnace with coke to reduce the oxide to lead bullion with a purity of about 97 per cent. The bullion is further refined by removing other metal impurities by skimming, after the addition of various reactive chemicals.

Major uses

There is a doubtful future for the use of lead in three important areas. Its most important use in car lead/acid batteries may be short-lived as most authorities believe that lighter, more efficient batteries will replace the lead battery within a decade or so. Its second major use as an additive to petrol is coming under greater scrutiny due to pollution problems. A minor, but still important use in type metal for printing is fast becoming obsolete.

Lead uses for making solders, ammunition and anti-corrosive sheaths for electric cables, however, are still strong markets. Other important minor uses for lead are in anti-radiation shields and as salts in the chemical industry.

Main market features

Lead is currently supplied mainly from developed countries, notably the United States, but futures reserves are likely to come from developing countries. Lead has been traditionally mined, and smelted with its 'sister' metal, zinc, from the same ore bodies. But these metals have different outlets, and an increasing amount of lead is now derived from scrap. The reclamation of lead from batteries and other sources has been highly developed and a good quality grade, competitive with many primary lead outlets, is obtained. An important influence in the lead market is that the growth in demand tends to be steady rather than spectacular. This applies particularly to the batteries market.

In times of economic crisis, batteries are still used by car owners and over the years it has been found that as each use for lead diminishes, such as in pipes and cables, it is replaced by another outlet.

Known world reserves

95 million metric tons

Methods of marketing and pricing

Lead is marketed in a similar manner to copper, which also has a large source of scrap supplies. In North America lead producers operate a fixed producer price

108

system, which applies to both sales of refined lead and as the basis for the cost of ore concentrates bought by refiners. Outside North America the bulk of lead is sold under direct supply contracts between producers and consumers, with prices based on the London Metal Exchange daily quotations.

Since primary lead producers are relatively few in number, and concentrated mainly in developed countries, they have, in the past, exercised considerable influence over the London Metal Exchange market, often keeping up prices by support-buying. In recent years, the threat of anti-cartel action and a large surplus of supplies has diminished this cooperative action by producers. At the same time producers are experiencing greater difficulty in controlling the Metal Exchange price movements in view of the large and increasing proportion of available scrap supplies that are generated if prices are pushed too high. The Metal Exchange itself acts as an important market of last resort for either buyers wanting supplies or sellers disposing of surpluses. In recent years it has been widely used for both these purposes by the communist bloc countries, notably the Soviet Union.

Lead prices

£ per metric ton. LME quotation

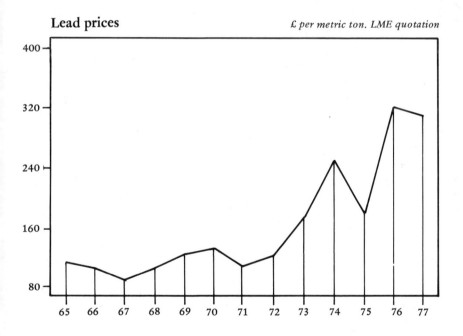

Lithium

Lithium ore production 1976

metric tons x 1,000

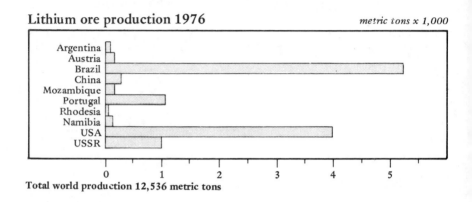

Total world production 12,536 metric tons

Grades available

Pure lithium metal is seldom used or traded. It reacts violently with water so must be kept in strictly dry conditions.

Most applications are for lithium compounds, the most common being lithium carbonate. Other useful compounds are hydroxide, chloride and fluoride.

Production method

Lithium's major ores are petalite, spodumene and lepidolite. It is also found in recoverable quantities in certain brines. The compounds are recovered from the ore by chemical means, either by an acid or alkaline treatment.

These are recovered from brines by a leaching and flotation method after concentration.

Major uses

Lithium is the lightest of all metals, but its uses are limited because of its reaction to water. Small quantities of the metal are used for making light weight alloys in combination with magnesium and aluminium. It is also used as a scavenger and degasifier in the production of steel and as a deoxidizer in melting copper and copper alloys.

The main uses for its compounds are as follows:-

a) *Lithium carbonate* This is used for ceramics, particularly enamels. A comparatively recent but major use is in aluminium smelting where it is added to the molten alumina. Here it has the effect of reducing the temperature at which electrolysis takes place and improves the yield of aluminium production.

b) *Lithium hydroxide* The major use for this compound is in the production of greases.

Lithium compounds are also used in the manufacture of bleaches, disinfectants, synthetic rubber, in welding, brazing and in the glass industries.

Main market features

The only major producers of lithium and lithium compounds are the US and USSR and to a certain extent West Germany. The two major US producers effectively control prices for the Western world but they have to compete with Russian exports outside the US.

There are signs, however, that the USSR is consuming more of its own production lately, thus reducing competition for Western producers.

The US has huge reserves of lithium ore and lithium containing brines, and is not dependent on imports of raw materials.

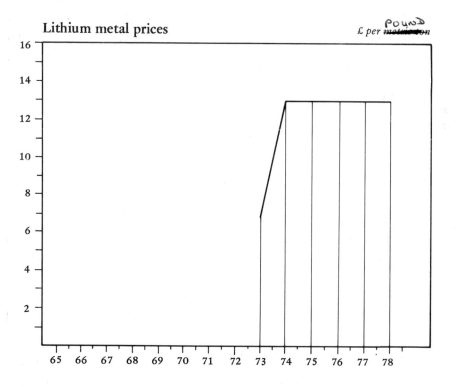

Lithium metal prices   £ per Pound

Reliance on imports of feedstock would limit future production outside lithium ore producing countries.

Known world reserves

6,036,000 tons, of which approximately 4,000,000 are estimated to be in the US.

Method of marketing and pricing

Merchant activity is small and confined to the marketing of USSR material, but the Russians are only intermittent sellers on the world market, so most consumers only buy Russian material to 'top up' supplies.

There is a small amount of competition between Western producers but their prices are usually very similar.

Magnesium

Magnesium production 1976

metric tons x 1,000

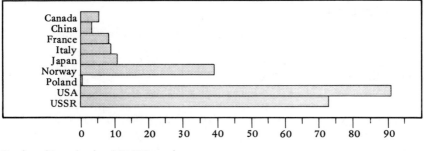

Total world production 240,500 metric tons

Grades available

Magnesium is traded in the form of ingots usually between 5-20kg each with the minimum purity of 99.8 per cent or 99.9 per cent.

Production method

Magnesium is found naturally in the ores — magnesite and dolomite, the latter being its principal source in the earlier days of its commercial exploitation. It is now more commonly extracted from brines in salt lakes or seawater. This innovation is likely to represent almost all future production.

Extraction from the ore The ore is first crushed, then thermally reduced with ferro-silicon at high temperatures in retorts. Magnesium comes over as a vapour which is then condensed.

Extraction from brines (magnesium chloride) The chloride is melted in a steel electrolytic cell into which graphite anodes are suspended. The magnesium is released by the passage of a direct electrical current and rises to the surface of the melt where it is tapped into ingot moulds.

Major uses

Magnesium is a metal similar to aluminium, in that it is light and its alloys are strong. It also has many of the same applications.

Magnesium is used in alloy form to make castings, particularly for the aircraft and automobile industries. It is also used in small proportions as an alloy addition in certain aluminium alloys: one of these alloys is used to make the ring-pull type can tops.

Another very important use is in the refining of other metals, notably titanium, where it is used to reduce titanium tetrachloride to the metal.

Ferro-magnesium, usually in the form of ferro-silicon-magnesium, is used in the manufacture of certain cast irons.

Main market features

Magnesium is approximately one third less dense than its rival, aluminium, and its alloys are almost comparable in strength. Theoretically this makes the metal more useful than aluminium for many applications, particularly in transportation.

The technology for the production and working of magnesium has been less advanced than that for aluminium, however, and many potential users have been concerned with the dangers involved in handling the molten or finely divided metal — both can oxidize violently in air. New fluxes and chemical covers can, however, minimize these dangers to a level comparable with those encountered in using aluminium. This should encourage more applications.

World production of magnesium is less than one twentieth of that of aluminium, which automatically increases its unit cost. It is this higher cost which has been the main reason for its comparative lack of development. However, consumption has been growing at a comparable percentage rate with that of aluminium. Magnesium manufacturers have long predicted a future boom in magnesium production once the production grows large enough to reduce unit cost.

Unfortunately, there are very few magnesium producers, partly for the reasons outlined above and partly because production should ideally be sited in an area of low electricity costs. With the advent of production from seawater, however, more sites may be considered suitable.

Prices cannot be expected to rise substantially unless the aluminium price rises disproportionately or unless titanium production is drastically increased.

Known world reserves

Reserves of magnesium are limited only by man's ability to extract it from seawater.

Method of marketing and pricing

With the exception of the USSR, production is confined to industrialized

Western countries, which have so far managed to retain a tight grip on the price, particularly as it applies in the country of production.

Merchant activity is almost exclusively confined to the trading of Russian material. USSR exports are not expected to keep pace with demand from Western countries, however, as they are likely to require more material for internal consumption. If this happens, free market activity will become further reduced.

Magnesium metal prices

£ per metric ton

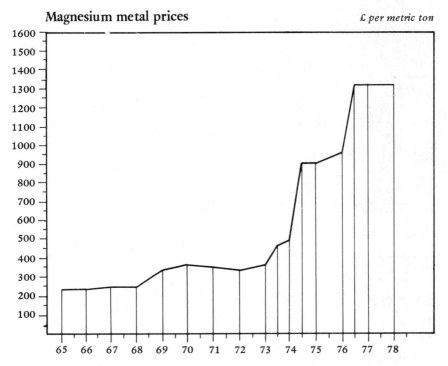

Manganese

Manganese ore production 1976

metric tons x 100,000

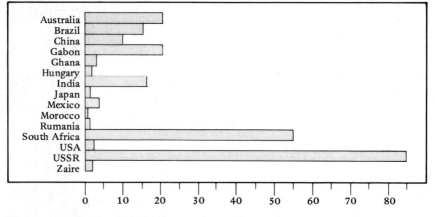

Total world production 24,100,000

Grades available

Pure manganese, usually in the form of an electrolytic lump, is available with a purity of 99.8 per cent minimum or 99.9 per cent minimum. Other forms of manganese on the market are ferro-manganese, with up to 80 percent Mn content, and silico-manganese, with a content of up to 20 per cent Si, balance Mn. Prices for ferro-manganese vary with the Mn content and carbon contamination.

Production method

Manganese is found naturally in the form of a number of different ores. The ores are concentrated by normal methods which vary according to the type of ore.

Ferro-manganese is produced by smelting the concentrate in either a blast furnace or an electric arc furnace.

Major uses

About three quarters of all manganese produced is consumed in the form of ferro-manganese used in the manufacture of manganese-containing irons and steels. These are in turn used in a wide variety of applications covering the entire spectrum of industry from construction steels used for bridges, to the steel used for home appliances.

Small proportions of pure manganese are used as an addition to certain aluminium alloys.

Manganese compounds are used in the manufacture of dry cell batteries, in the manufacture of some dyes and bleaches and also to produce several other important chemicals.

Main market features

Although the USSR and South Africa together produce about half the world's manganese (a feature shared with several other metals) there are a number of other important producing countries. Ferro-manganese is produced in an even larger number of countries. It is very difficult for any one producer to control prices, especially as manganese is such a large market.

There is no strict relationship between the price of the ore and ferro-manganese because of the varying cost of production, but because demand for manganese accurately reflects the world's general economic climate, the prices of both have moved historically in unison.

The market in individual consuming countries can be confused with changing import tariffs.

Known world reserves

Reserves of manganese contained in ore are approximately 728 million metric tons. This does not include manganese which may be found on the ocean floor in the form of nodules.

Method of marketing and pricing

Most ferro-manganese is consumed by large companies (iron and steel producers). They usually buy their supplies on long-term contracts directly from producers or producers' agents, or from a merchant, who has in turn bought long-term supplies from the producer. The reason for this style of business is understandable when one considers that ferro-manganese is a very cheap commodity traded in comparatively large tonnages. In order to make prices competitive, material must be shipped directly from producers to the consumers works. Profit margins per metric ton are generally so small that if material has to be stored awaiting a buyer, or shipped to a port further away than the one originally intended, the extra costs would not be met by profit margins.

The largest producers sell at a producer price but smaller producers and merchants trade on a free market basis at discounts or premiums to the producer price, depending on the state of the market.

There is a small, but active, market in ferro-manganese between merchants and smaller consumers such as cast-iron foundries.

The manganese metal market is not particularly interesting. The metal is consumed mainly by secondary aluminium alloy makers (those utilizing aluminium scrap in their feedstock). There are few large companies of this type, so their intake of manganese is usually only a few metric tons per month. There are merchants who specialize in this type of business, but again they have to compete with producers who sell to the consumers either directly or through agents.

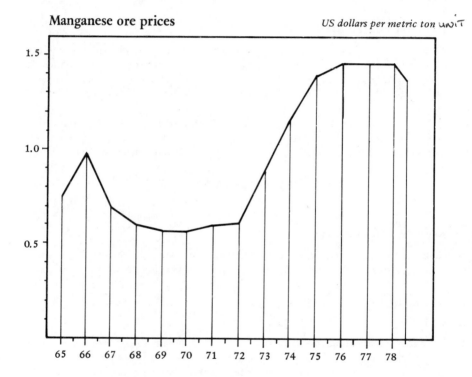

Manganese ore prices   *US dollars per metric ton* unit

Mercury

Mercury production 1975

flasks (76 pounds each) x 1,000

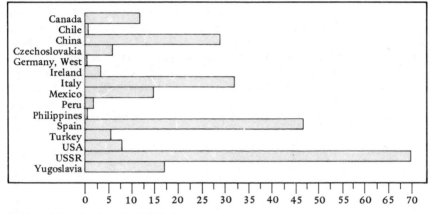

Total world production 273,739 flasks

Grades available

Mercury is sold by the flask, the steel bottle containing 76 lbs of the liquid metal. The standard purity is 99.99 per cent minimum.

Production method

Mercury is produced from its naturally occurring sulphide ore, cinnabar.

The ore is crushed but no attempt is made at further concentration. It is then roasted in retorts and the mercury metal is vapourized and collected. It is a very simple process. In some areas of the world mines produce only two or three flasks per week and are worked by only one or two miners who both excavate and refine the ore.

Major uses

Mercury is an extremely toxic metal. All its uses have come under increasing scrutiny because of its possible effect on the environment.

The best known use for mercury is in thermometers, barometers and pressure gauges, but this accounts for only about 10 per cent of its consumption.

It has three major uses. The first is in the manufacture of chlorine from brine, but other methods can be used to produce chlorine and these are gaining favour for environmental reasons. The second is in batteries, but other types of batteries provide adequate alternatives. The third major use is in agriculture as pesticides and fungicides, but here too environmental considerations have reduced demand and encouraged the use of alternative substances.

Strangely, perhaps, mercury has several important uses in medicine and in dental alloys.

Main market features

Known reserves of mercury are rather small and led to speculation in the 1960's that mercury would become scarce. Such speculation, however, did not predict the recent international concern about pollution. These concerns were reinforced by evidence that mercury has caused many deaths and much disease, the more serious of which are a result of chemical salts being dumped at sea or in rivers and then digested by fish. The mercury accumulates in the fish's body and is thus passed on to humans.

Mercury is the only metal which is in liquid form at normal temperatures and

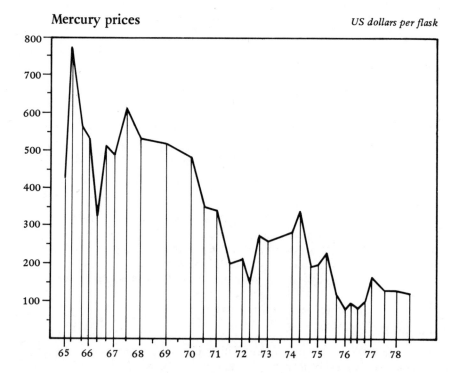

Mercury prices   *US dollars per flask*

will always find a use for that reason, but where alternative materials can be used they will be used. And these alternative materials probably account for over half the world's current consumption. This notion combined with very low prices has caused most producers, particularly in Spain and Italy, drastically to curtail production and sales. These moves have not improved the price.

Known world reserves

Commercially viable ore contains approximately 3,200,000 flasks.

Method of marketing and pricing

There are no effective international producer prices for mercury. All producers have to sell at a free market price, but many large orders for the metal are purchased on a tender basis, where merchants and producers alike submit offers and the cheapest offer wins the contract. In the depressed circumstances of the mercury market consumers have the upper hand.

In the 1960's there was a very active and important market in mercury which attracted many large trading houses. With the combination of low prices, low turnover and the absence of price movements, there are now few merchants specializing in this trade.

Scrap recovery

There is a fair trade in scrap mercury, usually carried out on a local basis. This material can be cleaned by filtration or by distillation.

Molybdenum

Molybdenum ore production 1976 *metric tons x 1,000 M_0S_2 content*

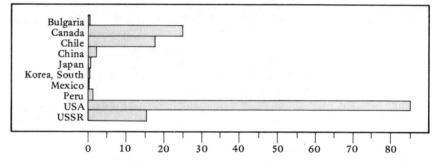

Total world production 148,000

Grades available

There is only a small world consumption of molybdenum in its metallic form. The bulk of the metal is traded in the form of molybdic oxide (MoO3) or ferro-molybdenum, the alloy of molybdenum and iron, containing approximately between 60-80 per cent molybdenum or in the form of its concentrated ore (sulphide).

The few producers of the pure metal supply it in the form of ingots, sheet and wire.

Production method

The naturally occurring ore, molybdenite, a molybdenum sulphide, is concentrated by flotation. The oxide is produced by heating the sulphide in air. The oxide can then be purified by sublimation. The material is produced by the reduction of the pure oxide in the presence of hydrogen at high temperatures. Between one quarter and one third of the world's production of molybdenum is produced as a by-product of copper production when the copper ore contains a high proportion of molybdenum.

Major uses

The pure metal has a small but increasing application in highly technical industries such as nuclear energy, aerospace and electronics.

Its main application by far is as an alloy addition to steel. Its effect on steel in additions of about 1 per cent is much like that of chromium and nickel, it increases strength and toughness especially at high temperatures. Such steels find applications in the production of mining equipment, machine components and have a growing use in oil pipelines.

Major market features

The US is much the largest producer of molybdenum and it is produced there by only a few companies. These companies are therefore in a powerful position to control the market price. They publish a producer price for their major products which is adhered to but they ration material to their customers in times of shortage. Supply is somewhat affected by fluctuations in copper production but strangely, not much affected by the steel market variations due to the special applications of molybdenum steels, the uses of which may be somewhat out of phase with general economic conditions. The development of the oil industry is particularly relevant to demand due to its application in pipelines and refineries.

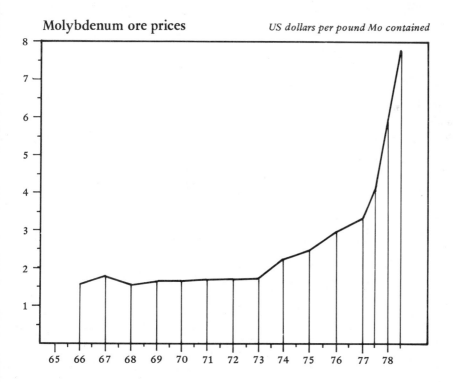

Molybdenum ore prices   *US dollars per pound Mo contained*

Average annual demand is expected to rise by between 3.5 per cent and 4.5 per cent per year. Current indications, however, may put these forecasts on the low side due to the increasing demand from the Soviet Union.

Supply is not expected to fall far behind demand unless copper production is drastically cut. Current demand trends indicate steadily rising prices.

Known world reserves

Approximately 4,900,000 metric tons of which 3 million metric tons are estimated to be in the United States.

Method of marketing and pricing

Molybdenum prices in the form of sulphide, oxide and ferro-molybdenum are quoted as a price per pound or per kilogram of molybdenum content.

Free market trade is somewhat intermittent, drying to a trickle in times of supply/demand balance but flaring up into hectic trading in times of shortage. This trade is conducted by comparatively few international trading houses which usually specialize in all the refractory metals that is vanadium, columbium, titanium and tungsten.

Nickel

Nickel production 1977

metric tons x 10,000

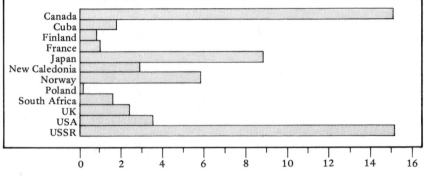

Total world production 704,400 metric tons

Grades available

Nickel comes in many shapes, sizes and purities to suit the various uses to which it is put.

Cathodes These are produced by electrolytic means and are either full size (un-cut) up to 750mm x 750mm or cut into smaller squares for convenience. These squares range from 1in x 1in up to 10ins x 10ins. The full size cathodes are cut into squares using a power shear, so they are only roughly square. The smaller squares are always more expensive than the larger ones as more cuts have to be made. As a very rough guide, 10ins squares are about two US cents per pound more than un-cut cathodes, but 1in x 1in squares may be 10 US cents per pound or more above the price of 10in x 10in squares. The thickness of cathodes varies with the producer, but generally is about 1cm. The smaller squares may be used for electroplating.

Minimum purity is usually 99.8 per cent with a maximum cobalt content of 0.1 per cent.

Briquettes These are small, regular shaped lumps of compressed nickel matte (spongy granules of pure nickel). They can be a variety of shapes — tablet, round or square, but generally no bigger than 2½ins in the largest dimension.

125

Granules These are solid pure nickel in granular form.

Pellets and Shot These are made only by the largest producers.

Ferro-nickel An alloy of nickel and iron usually containing about 30 per cent nickel.

Nickel oxide This is known sometimes as sinter, usually in briquetted form.

These are the main types of nickel traded. There are premiums and discounts depending on the type of material: cathodes and pure pellets are the most expensive and ferro-nickel the cheapest. All prices move up and down in unison. To complicate the situation further, some origins of metal may be more expensive than others. The best guide to price movement is the market for Russian 10ins x 10ins cathodes which are probably the most commonly traded on the free market.

Production method

Almost all nickel is mined as nickel ore rather than as a by-product of other metal ores. The ore is either an oxide or a sulphide of nickel. The sulphide ores are crushed and concentrated by flotation but with laterite (oxide ore) no concentration method has been found effective.

Where nickel is found in conjunction with copper, the nickel may be separated by a combination of magnetic and flotation techniques and by recrystallization. Sulphides are converted to oxides by roasting in air and the oxides reduced to the metal in a reverberatory furnace. Where nickel ore contains cobalt or other secondary metals separation is by chemical means.

Major uses

Nickel has a vast number of uses throughout industry. The most commonly known use must be in coins when alloyed with copper, and in stainless steel when alloyed with iron and chromium. It is used in metallic and salt form with and without other metals and salts, as a catalyst in a large range of inorganic chemical reactions. Many types of steel alloys contain nickel especially when resistance to corrosion is required. It is used for plating, especially in the automobile industry and for cutlery. Some types of batteries use nickel. It forms part of many high strength alloys used in the aircraft industry. Stainless steel can be made with several other metals but nickel is by far the most competitive in cost.

Main market features

Historically nickel has been produced by a limited number of powerful groups led by International Nickel of Canada, who at one stage had over 80 per cent of the western world market. However, the last 10 years have seen dramatic changes in the structure of the industry with the entry of many newcomers and International Nickel's share has been cut down to below 35 per cent. The turnround started in 1969 when a long strike at the Canadian mines of Inco, and Falconbridge, resulted in a severe shortage of supplies. This forced consumers to look for alternative sources, or substitutes for nickel, and attracted new producers into what then became a highly competitive market. Canadian sulphide deposits remain the main source of nickel at present, but the bulk of future reserves are in laterite deposits.

Method of marketing and pricing

The largest producers have maintained a producer price at most times but don't strictly adhere to this price, especially when contracting with the larger consumers. However, in 1977 INCO abandoned its role as price leader by witholding its producer quotation in favour of 'confidential' prices. There is an extremely active free market conducted by merchants and small producers and in the past this market has been both volatile and speculative.

Nickel prices

US dollars per pound. Free market.

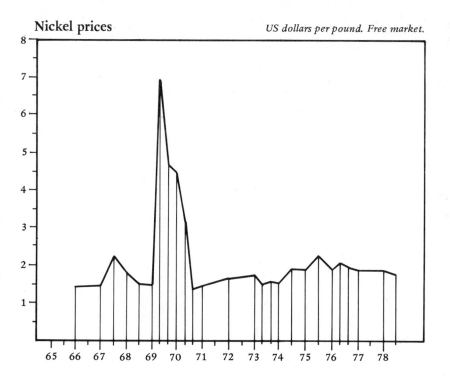

Known world reserves

Approximately 90 million metric tons.

Selenium

Selenium production 1975

kilograms x 10,000

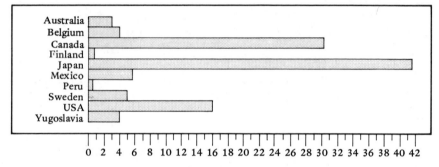

| | 0 2 4 6 8 10 12 14 16 18 20 22 24 26 28 30 32 34 36 38 40 42 |

Total world production 1,100,000 kilos

Grades available

Selenium is usually made available in the form of powder or granules and most commonly traded as a fine powder (minus 200 mesh) with a purity of either 99.5 per cent or 99.8 per cent minimum.

Production method

Almost all selenium is produced as a by-product of copper production. Selenium-rich slimes are recovered from the anode during the electrolytic refining of copper. To produce the elemental selenium these slimes are treated with sulphuric acid and the metal is then precipitated by a simple chemical operation.

Major uses

The five major uses of selenium are rather sophisticated and it should be noted that in each, the cost of the metal is small compared with the system or process in which it is used. These uses divide the consumption roughly equally but change slightly according to technical innovation.

It is used in the manufacture of certain types of glass as a decolourizer. In

128

the electronics industry it is used in the manufacture of transformers, semi-conductors and photo-electric cells. Selenium has a unique use in photocopying machines, transferring a photographic image by means of static electricity. The metal is used in combination with cadmium to make an orange/red pigment used extensively in plastics and ceramics. The remaining major use is as an addition to steels, mainly stainless steel, to improve forging characteristics. There are several very minor uses for selenium in the chemical, pharmaceutical, rubber and explosives industries.

Main market features

Selenium has one of the world's most volatile markets. It has all the classic ingredients of volatility: it has a small production, which makes it suitable for market manipulation; the demand for selenium is only fractionally reduced by high prices as explained above; and it is a by-product which sets an upper limit on production.

In times of shortage its value can change by hundreds of per cent only to fall by similar values when consumption declines. Fortunately its production is rather widespread compared with many materials, and no single producer is in a position completely to control prices. Demand in all its major uses is expected to remain strong with the possible exception of pigments due to the continuing

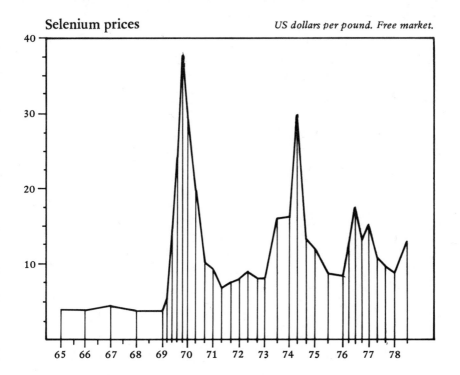

Selenium prices　　　　　　　　　*US dollars per pound. Free market.*

concern over heavy metal toxicity.

There is no doubt that the selenium market will continue to be extremely volatile in the future with a gradually rising average price.

Known world reserves

Approximately 80,000 tons.

Method of marketing and pricing

North American producers publish a producer price for the metal which they adhere to fairly strictly. Other producers may base supply contracts on this producer price but are increasingly marketing on the basis of free market quotations. This gives them the obvious advantage of making very large profits from the metal in times of shortage and prevents them from having to keep stocks of material in times of over-supply, as the North Americans are sometimes obliged to do, to protect their price.

The free market is fed particularly by material refined in Japan. International trading is conducted by merchants mainly in London, Dusseldorf and Brussels.

Silicon

Silicon production 1974

 *metric tons* **X 1,000**

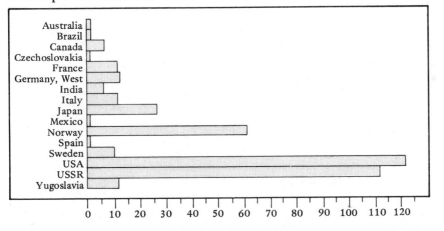

Total world production 455,000 metric tons

Grades available

Most silicon is marketed in the form of irregular lumps usually between 10-100mm in size. The most commonly accepted purity is 98.5 per cent minimum with maximum iron and calcium contents of 0.5 per cent and 0.3 per cent respectively. When silicon is used in the chemical industry, higher purities are often required.

Most silicon is consumed in the form of ferro-silicon an alloy of silicon and iron, the most popular grade contains 75 per cent silicon.

Production method

High grade quartzite sands and pebbles are the raw material for both silicon metal and ferro-silicon. These are mixed with coke in the production of silicon and with both coke and scrap iron in the production of ferro-silicon, and then smelted in a powerful electrical arc furnace. The raw materials for both products are extremely cheap. The cost of production is mainly the cost of electrical energy. Production plants are usually sited in areas with low cost electricity and often linked to large hydro-electric plants.

Major uses

Approximately 90 per cent of all silicon is produced in the form of ferro-silicon which is consumed in the iron and steel industry as an alloying agent. Consumption is divided roughly equally between steel and cast iron manufacturing.

About 75 per cent of all silicon metal produced is consumed in the aluminium industry where it is used to make alunimium alloys of a type that is used in castings.

Silicon metal (silicon is not strictly a metal in chemical terms and is sometimes referred to as a metaloid) is also used as a de-oxidizer in steel production and in the manufacture of silicones used as waxes, lubricants and polishes.

Main market features

Although we are concerned here with the marketing of pure metals rather than alloys, we must in this case take the ferro-silicon market into account because many plants producing ferro-silicon can be adapted to producing silicon metal and vice versa.

Aluminium, iron and steel have such wide applications that the markets of both metals reflect the general state of world economic conditions. As nearly all silicon is used in these two industries it follows that the silicon market is governed in the same way. A higher growth rate for aluminium over steel is predicted which will encourage silicon rather than ferro-silicon production, but silicon production requires more energy to produce than ferro-silicon plus a higher grade of raw material.

Many countries with spare electric power have been attracted to silicon production because the raw material is cheap and relatively common. Electricity is often a difficult commodity to export and production plants are fairly unsophisticated. These countries assume that silicon is like a solid storable exportable form of electrical power. The recent history of the market, however, shows great volatility and nothing like the comparative strength of the energy market. This volatility is caused by silicon production capacity being out of phase with the consumption of steel and aluminium.

Silicon is an extremely useful material and has a particularly important role in the production of pipelines and aluminium alloys both of which are high growth areas. This, combined with the continuing shortage of energy, is likely to produce a shortage of production capacity and a consequent increase in price, especially as prices in the mid-70s have discouraged the building of new plants.

Known world reserves

Silicon is one of the world's most common elements. Reserves are so enormous that there is no point in estimating them.

Method of marketing and pricing

Ferro-silicon, together with ferro-manganese and ferro-chrome, are known as bulk ferro-alloys and they are traded quite differently from silicon metal. Ferro-silicon is marketed to steel mills in very large tonnages often more than 1,000 tons at a time, either directly by the producers or very often through their agents. There are comparatively few steel mills, so trade is highly competitive, except in times of shortage. There are, however, very many cast iron foundries which buy in quite small quantities from producers' agents and merchants who specialize in the trading and distribution of bulk ferro-alloys. Most of these merchants are small, nationally based companies, but there are several very large international merchants trading in larger tonnages who often have agency agreements with the producers.

There are comparatively loose forms of producer price control but most trade is subject to supply and demand.

Silicon metal is also distributed directly by producers or their agents at a ruling producer price, but there is a large and active free market in the

Silicon prices

US dollars per metric ton. Free market.

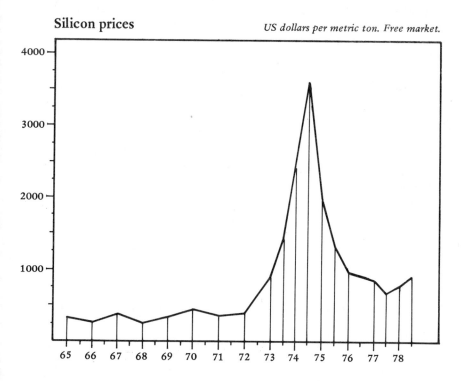

metal. The Scandinavian producers are the most successful at maintaining a producer price. Merchants in silicon metal are usually different from those trading in ferro-silicon. The material they deal in comes mainly from Yugoslavia and Spain, and from that which is offered for export by producers outside the country of production.

Sodium

Grades available

Sodium is an extremely soft metal which reacts violently with water. It is usually sold in the form of sticks or blocks which have to be protected from water in the atmosphere by submerging in oil or by packing in airtight drums.

Production method

Sodium is produced electrolytically from molten sodium chloride (common salt).

Major uses

About 80 per cent of production is used to make the anti-knocking agents tetraethyl and tetramethyl lead used in petrol. Its other major uses are in the manufacture of titanium metal (magnesium metal can also be used for this purpose) and as a coolant in nuclear reactors.

Main market features

There is very little merchant activity in sodium. This is mainly because there are few producers in the world, and the main producers also manufacture the anti-knocking compounds, which are the biggest outlet for this metal. The USSR does export material from time to time but only when in surplus to its own requirements.

Known world reserves

Reserves of raw material are only as limited as the salt in the ocean.

Method of marketing and pricing

The market price of sodium is based only on its cost price. The raw material costs next to nothing and the main costs of production are electricity and handling, which have been rising rapidly. Competition is not particularly fierce for the reasons outlined above and because of the danger of moving the material over long distances, or storing for long periods of time. All major consumers buy direct from producers who sell at a similar price in the same market.

Tantalum

Tantalum (concentrates) production 1974 *metric tons x 100*

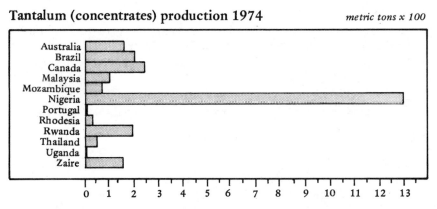

Total world production 2,545 metric tons

Grades available

There is very little free trade in tantalum, but the producers' main products are powder, sheet and bar, 99.9 per cent minimum purity. Merchant activity is restricted to scrap solids (sheet and bar off-cuts).

Production methods

Tantalum oxide is extracted from its natural occurring ores, tantalite and columbite (the two often occur together). The oxide is first concentrated by magnetic separation and the metal is then produced by electrolytic means.

Major uses

Like tungsten and columbium, tantalum is one of the refractory metals. It has a high melting point and is very resistant to corrosion. In the past it was added to steel to make alloys similar to those made with columbium. The discovery of adequate deposits of columbium ore which is less dense than tantalum has almost eliminated the use of tantalum as a steel additive.

Tantalum's corrosion resistance has found many applications in the chemical industry where it is used to make pipes, crucibles, retorts, etc.

The electronics industry is tantalum's biggest consumer. Here it is used to manufacture components such as contact points, valves and electrodes. Small quantities of tantalum carbide are used in cutting tools and even smaller quantities are used in surgery to make clips and struts to repair veins and bones.

Main market features

Alternative materials exist for all the major uses of tantalum which sets a simple economic limit on its price. The metal is also quite expensive to produce, which confirms a lower limit.

Recent increases in demand for the other refractory metals with their wider applications has had the effect of tightening supplies and therefore raising the price. This has enabled tantalum producers to keep pace with rising production costs and this situation is likely to be maintained for the foreseeable future.

Tantalum ore is produced in too many countries to encourage any attempt at cartels or price control.

Method of marketing and pricing

Tantalum ore has an important application for making refractories (used for

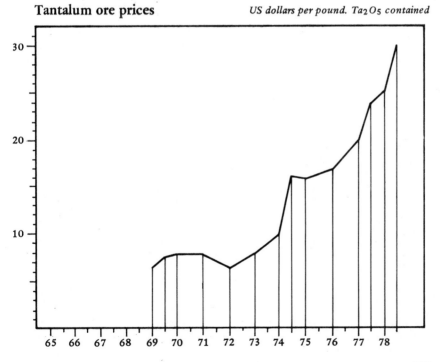

Tantalum ore prices   *US dollars per pound. Ta_2O_5 contained*

furnace linings, etc) outside its use as a raw material for making tantalum. Supplies are adequate, however, and there is no likelihood of there being restrictions on tantalum metal production because of shortages of ore, especially since such a small proportion of the ore mined is used to make the metal.

All tantalum metal producers maintain a close relationship with consumers and are often required to make the semi-fabricated parts that consumers require. These products may have a strict degree of technical tolerance which is reflected in the price.

Such a trade doesn't lend itself to merchant activity but a few merchants are involved in trading scrap tantalum which is usually recycled back to the producers.

Tellurium

Tellurium metal production 1975

kilos x 1,000

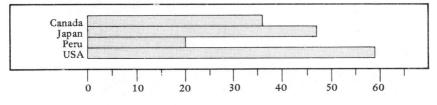

```
                0    10    20    30    40    50    60
Canada
Japan
Peru
USA
```

Total world consumption 163,287

Grades available

Tellurium is usually marketed in the form of small ingots or slabs, or in the form of fine powder. The purity of commercial grade material is normally between 99.7 per cent and 99.9 per cent minimum.

Production method

Tellurium is only produced as a by-product of other metals, particularly copper, but it may be present in commercially recoverable quantities in certain zinc and lead ores. Tellurium containing slimes are recovered from the anode during the electrolytic refining of copper. These slimes are leached with caustic soda. Tellurium oxide is then precipitated from the solution by acidification and the oxide is roasted to the metal and refined.

Major uses

Tellurium is used as an alloy addition, in very small proportions to certain steels and copper alloys. In both cases the tellurium content improves the machinability of the alloy. It may be added to iron and steel in the form of ferro-tellurium containing about 50 per cent tellurium, the balance is iron.

Its addition to steels reduces porosity and acts as a grain refiner.

The compounds have various uses in the chemical industry in ceramics, rubber vulcanization and in detonators and pesticides. However, these chemical uses only amount to about 10 per cent of the tellurium market.

139

An alloy of bismuth and tellurium displays unique electro-thermic characteristics, but no major industrial application has yet been found for this property.

Main market features

Tellurium's uses are of no particular importance and all have known substitutes, but tellurium is preferred to other products at current commercial differentials.

Only about half of the potential supply of tellurium in copper ore is utilized at present, due to the cost of its extraction. Any future increase in demand could be met quite easily by improving extraction techniques.

Prices are likely to keep in line with the increased cost of production, but any future growth in demand may lead to a temporary price rise while production is being increased.

Known world reserves

Approximately 50,000 tons of tellurium metal.

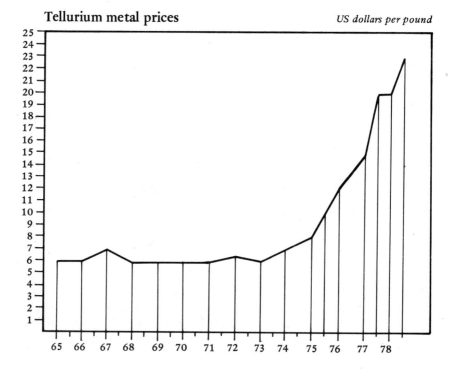

Tellurium metal prices
US dollars per pound

Method of marketing and pricing

The tellurium market is very small and its supply/demand characteristics have not attracted much merchant activity. The major producers handle most of the marketing directly or through agents. Prices are fixed by producers, but some business is done on a free market basis, especially that between producers and state-owned buying corporations, often with a merchant as intermediary.

Tin

Tin mine production 1976

metric tons x 1,000

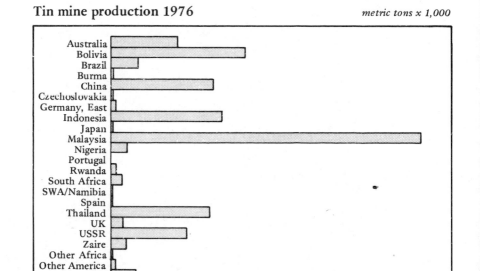

Total world production 216,400 metric tons

Grades available

Tin is sold on the international market in ingot form (up to 50kg each) with a purity of 99.75 per cent or at a premium price for high grade material with a purity of 99.85 per cent minimum.

Production method

Tin's main ore, cassiterite (tin oxide), is mainly found in alluvial deposits in river beds and deltas and recovered by dredging methods. It is also recovered by the gravel pump mining method, and by traditional mining in such areas as Bolivia.

The ore, either from dry or wet deposits, is usually concentrated by magnetic or electrostatic means and the concentrate is smelted with coke in a reverberatory furnace to reduce it to metal.

Major uses

The major user of tin is still the canning industry. Tin cans, used for packaging food and drink, are made of steel sheet, plated with a thin layer of tin. This material is known as tinplate which is rolled from a large slab of steel clad with tin. Tin is an important constituent of most solders and other alloys such as babbit metal and bronze.

Main market features

Tin is found in relatively few countries. Malaysia is by far the biggest producer, followed by Bolivia, Indonesia, Thailand, China, the Soviet Union, Australia, Brazil, Nigeria, Zaire and the United Kingdom. It is one of the few metals found only in limited quantities on the North American continent. But the United States holds a strategic stockpile, equal to about one year's total world consumption. No major new deposits of tin have been discovered for many years, but the development of offshore mining has to an extent offset the gradual exhaustion of land deposits. Improved mining techniques, and higher price levels, have also made practical the exploitation of deposits with a far lower tin content and encouraged the reworking of 'tailings' from old mines. Over half the world's refined tin metal is smelted in Malaysia, at Penang, where the smelters produce the top quality Straits tin. Indonesia and Thailand have their own domestic smelters. Bolivia is building up its smelting facilities, but still exports a large proportion of its ore concentrates to smelters in Europe, notably the United Kingdom.

There is some reclamation of tin previously used, but since the main outlet is as a thin coating on steel to produce tinplate, scrap recovery is limited. Historically, demand for tin has had a very low annual growth rate which usually fluctuates in accordance with the state of the steel industry. There has been strong competition from so-called 'tin free' steel — that is tinplate with another metal such as chrome replacing tin — and from aluminium in the beverage cans market.

Known world reserves

6.5 million metric tons

Method of marketing and pricing

Tin is marketed in a variety of ways: by supply contracts between Governments or individual companies, with consumers; through merchants; or through the Penang smelters or London Metal Exchange. Another supply source is the United States stockpile, which sells by competitive bids.

The two main pricing centres are the Penang market in Malaysia which is the main physical market and the London Metal Exchange. The smelters daily offer for sale a quantity of tin equal to the intake of ore concentrates received from miners. Buyers put in their bids, and under a complicated formula the smelters calculate an average price for that day of the tin available for sale. They take the top price bids down to the point where all the tin available has been bidden for, and the bid, at that cut-off point, becomes the average price.

The London Metal Exchange tin market trades in tin for delivery up to three months ahead, although individual dealers will provide quotations for even further forward. It has the only futures 'hedging' mechanism which gives protection against price fluctuations. (The Malaysian Government is at present planning a tin futures market.) The London Metal Exchange warehouses attract supplies from all over the world, including Malaysia, so there is a physical trade business as well as a 'paper' futures market.

Prices for supply contracts are often based on a mixture of quotations from the Penang and London Metal Exchange, and there is a great deal of arbitrage between the two markets if one or other moves out of line.

The tin market has a long history of price volatility but in the past 20 years it has been controlled to a large extent by the International Tin Agreement, between producers and consumers. The agreement defines 'floor' and 'ceiling' price levels and attempts to keep prices between these limits by the use of a buffer stock mechanism, backed on occasions by export quotas. An important influence on the market has been the US strategic stockpile, built up after the Korean War, which still holds the equivalent of about a year's world

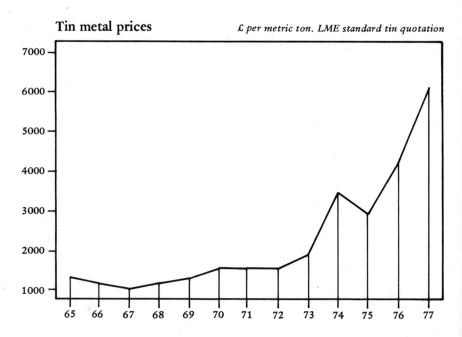

Tin metal prices   *£ per metric ton. LME standard tin quotation*

consumption. Stockpile releases have been used in the past to fill the shortfall between new production and demand. With the stockpile diminishing, and the US becoming a member of the Tin Agreement, which will have a much bigger buffer stock, the importance of the stockpile as a market influence could diminish.

Titanium

Titanium metal production 1976 *metric tons x 1,000*

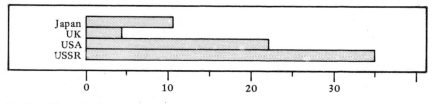

Total world production 73,250 metric tons

Grades available

Titanium's uses are of a highly technical nature. Many producers are therefore obliged to make titanium in several sizes and forms, that is slabs, sheet, bars, wire and other semi-fabricated types, but pure titanium is traded mainly in the form of sponge (irregular granules) or briquettes (compressed blocks of sponge). This material is usually of 99.6 per cent minimum purity.

Production method

The metal is produced by first converting the ore, rutile or ilmenite into titanium dioxide by means of a simple chemical treatment. The dioxide is then converted to the tetrachloride and heated with magnesium or sodium which reduces it to the metal.

Major uses

There are many independent industrial uses for both the ore and the oxide. Titanium metal's physical properties, its lightness, strength and resistance to corrosion have resulted in many new uses for the metal in modern times. It has been found particularly useful in the aerospace industry, where it is used for such items as turbine blades, engine shafts and in the framework and skins of spacecraft and high performance aircraft. Its resistance to corrosion makes the metal suitable for applications such as pipes, pumps and containers in the chemical industry and in electro-plating. The metal has found an increasing use as an alloy addition in certain steels, some of which are used in automobile

146

silencers (mufflers). It is added to steel either as pure titanium or in the form of ferro-titanium, a master alloy which is made by melting scrap titanium and iron, or by direct reduction of the ore with iron.

Main market features

Although titanium's applications in industry are constantly widening, its market is still very much dependent on the fortunes of the aerospace industry. The value of the metal bears almost no relationship to the price of the ore, but more to the value of the pure metal (magnesium or sodium) required to reduce it from the ore. Both these metals are in abundant supply in sea water but incur high energy costs in production.

Titanium production is presently in the hands of very few manufacturers in industrialized countries and it is unlikely that this situation will change in the foreseeable future.

Known world reserves

Titanium metal in ore is approximately 4,350,000 metric tons. However, titanium can be extracted from other minerals less competitively.

Titanium (sponge) prices   *US dollars per pound*

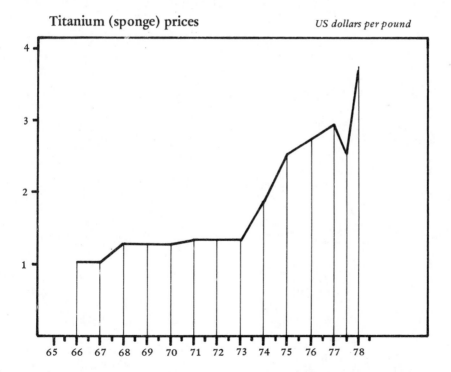

Method of marketing and pricing

The producers of titanium control its price in the country of production. Their price is limited only by the price of imported titanium products from other countries. There is, however, a small free market trade in metal originating from the USSR, usually in the form of sponge or briquettes. Small quantities are also available from Japan. Scrap titanium is a more interesting market however. This material arises in the form of machine turnings, sheet cuttings or solids. These come mainly from large aircraft manufacturing centres, such as the US, the USSR and West Germany. Prices vary according to the alloy type, form and contamination.

Tungsten

Tungsten ore production (WO$_3$ content) 1975 *metric tons x 1,000*

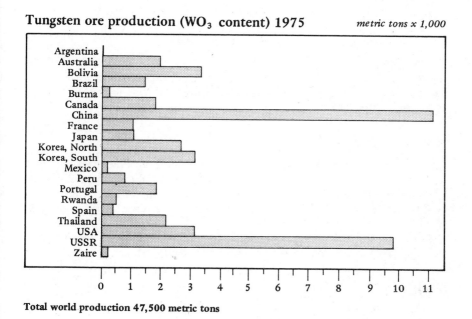

Total world production 47,500 metric tons

Grades available

Tungsten is traded mainly in the intermediate forms of the metal. These are the concentrates, wolframite and scheelite, ferro-tungsten (the alloy of iron and tungsten) and ammonium paratungstate (APT). Pure tungsten, usually in the form of powder, is also traded, but this is a very minor proportion of the business.

Up to 90 per cent of the world's international business in tungsten is conducted in the concentrates. These are traded in metric ton units. One MTU is a single per cent of WO$_3$ contained in a metric ton of the concentrate, i.e. 22,046 lbs of WO$_3$. Most commercial grades of tungsten concentrates contain about 65-70 per cent WO$_3$ which means that they have 65-70 MTU per metric ton.

149

The concentrates traded are scheelite ($CaWO_4$), ferberite ($FeWO_4$), huebnerite ($MnWO_4$) and wolframite ($[FeMn]WO_4$). Ferberite, huebnerite and are loosely referred to as wolfram ore, or black ore in the trade. Traders pay particular attention to the levels of impurities in these concentrates. The most important of these are tin, arsenic, phosphorus, sulphur, copper and molybdenum. The average scheelite or wolfram ore would contain:

WO_3 (tungsten oxide)	68 per cent
Sn (tin)	0.2 per cent maximum
As (arsenic)	0.2 per cent maximum
P (phosphorus)	0.04 per cent maximum
S (sulphur)	0.5 per cent maximum
Cu (copper)	0.1 per cent maximum
Mo (molybdenum)	0.3 per cent maximum

To confuse the issue further a product called synthetic scheelite is regularly traded which is chemically produced from low grade ores. It is less popular than natural scheelite due mainly to the higher import tariffs it carries into some consumer countries.

Production method

All tungsten concentrates are produced by very simple flotation and gravitational separation from the ore. Ferro-tungsten is either produced by the normal aluminothermic method (reduced from the ore with aluminium powder in the presence of iron) or by reduction in an electric arc furnace. Tungsten scrap is often the feedstock in this last method.

Most pure tungsten is produced in powder form from APT by reduction with hydrogen. APT is produced chemically from the concentrates.

Major uses

There are very many tungsten-containing alloys. Some are steels (scheelite can be used as a direct addition to steel for making many of these alloys) with wide application in all forms of machine tools and mining equipment. Tungsten carbide alloys, made by sintering tungsten powder with other metal powders, also have very wide applications for tools, machinery, components, bullets and tyre studs, etc.

Fine tungsten wire is used for electric lamp filaments, and tungsten metal or alloys are used as contact points in electrical switch gear.

Tungsten compounds are widely used in the chemical industry in lubricants, pigments, catalysts and in many other minor applications.

Main market features

Most of the uses for tungsten are vitally important for the world's industrial activity. Substitution on a large scale with other materials in these uses is very difficult. Fluctuations in demand usually reflect the general state of the world's economic situation, but the price trend in the last two decades has been rising substantially due mainly to the increase in demand and to growing control by tungsten producers.

Tungsten is a typical example of a vitally important raw material which is produced mainly in third world countries but consumed mainly in industrialized countries.

China alone produces over a quarter of the world's supply and is in a position to exercise a massive degree of control over prices. The world's producers display a degree of cooperation but the smaller producers in South East Asia and South America rely upon China to set the pace.

With almost no substitution possible and a rising demand, it is reasonable to suppose producers will exert control over supplies to ensure that prices continue to rise, especially as China is expected to increase its internal demand for tungsten over the next decade.

Known world reserves

Reserves of tungsten ore (wolfram content) are approximately 1,200,000 metric tons. 900,000 metric tons are estimated to be in China.

Method of marketing and pricing

The tungsten market, perhaps more than any other, is conducted by international merchants.

It is extremely volatile and very speculative. The purchases from China and the sales to Eastern European countries including the USSR, are the backbone of the trade. Very little trade is done directly between them for political and financial reasons.

The Chinese conduct a responsible, if somewhat erratic, sales policy. They may withdraw entirely as sellers for long periods of time if they feel the price is too low, only to appear as aggressive sellers when the price suits them or when they have a requirement for foreign currency. Their No 1 grade wolfram ore specification is a standard for world trade.

The other smaller producers have their own specification for their product, and assays usually have to be made by independent assayers on the outturn of cargoes.

With the exception of North America, very few producers sell directly to consumers.

The most authoritative price for the ore is recorded in the London Metal Bulletin, but there is much bitter controversy from producers, consumers and merchants alike concerning the accuracy of these records.

Many efforts have been made to find a more suitable way of international pricing but without success.

In 1978 an International Price Index, using information drawn from both consumers and western world producers was introduced as a means of reflecting more price influences than just the free market merchant price published by the Metal Bulletin.

151

Tungsten is an expensive product with a volatile market which has attracted intense speculation, often with dire results for the speculator. Without this speculation, however, it would be difficult to envisage how the tungsten market could be conducted.

The General Services Administration (GSA), the American strategic stockpile retains large stocks of tungsten in various forms. This material is currently made available to buyers in regular official sales. It is an additional confusing ingredient in this very complicated market.

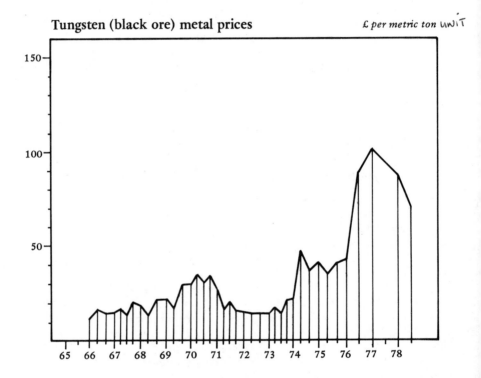

Tungsten (black ore) metal prices £ per metric ton UNiT

Vanadium

Vanadium ore production 1975

metric tons x 1,000

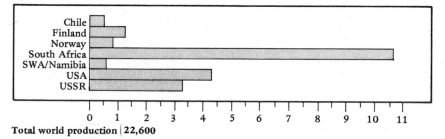

Total world production | 22,600

Grades available

The pure metal is seldom traded, as vanadium additions to steel are usually made in the form of ferro-vanadium. The bulk of trades are in vanadium pentoxide (V_2O_5) or ferro-vanadium usually containing either 80 per cent or 50 per cent vanadium, the balance is iron.

Production method

Although vanadium is a very common metal in the earth's crust, it is only rarely found in commercially viable concentrations often in association with other ores and minerals.

It is also extracted from the flue dust of power stations which use vanadium rich cokes and coals.

Ores or vanadium-containing slags are converted into vanadium oxide or directly into ferro-vanadium.

The raw material is concentrated, leached and roasted to the oxide.

Ferro-vanadium is usually produced aluminothermically by reduction of the oxide in the presence of aluminium powder and iron.

Major uses

Most vanadium is used in the form of ferro-vanadium in the production of structural steels. An important growth area for such steels is in the manufacture of oil and gas pipelines. The growth in production of pipelines has become particularly marked since the tremendous rise in fuel prices in 1974 and 1975.

There are important uses for such steels in bridge building and high-rise buildings.

Other types of steels containing vanadium are used wherever toughness is required in machinery components, mining equipment and tools.

Certain non-ferrous alloys incorporate a vanadium content, notably some titanium alloys.

Vanadium compounds are used in the chemical industry as catalysts in glazes and for the manufacture of some synthetic rubbers.

Main market features

South Africa produces approximately half the world's vanadium which implies that it should be in a position to exercise a degree of control on the market. Although it is true that the South Africans could set the pace for price changes, they appear to be conscious of the fact that the other so-called refractory metals, tungsten, molybdenum and niobium are able to replace vanadium in many of its applications.

This feature of the market perhaps more than any other has kept price movements reasonably gentle.

The current boom in pipeline construction has created an increased demand

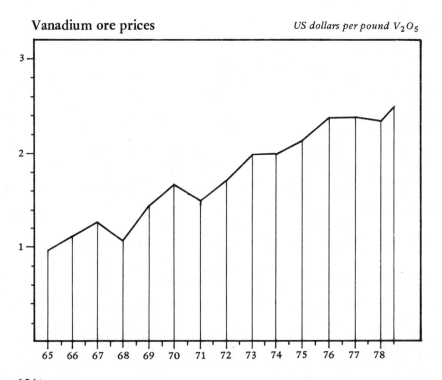

Vanadium ore prices   *US dollars per pound V_2O_5*

for vanadium and its competitors which may signal a change in this historic pattern. Any long-term shortage is likely to tempt the few major vanadium producing countries to flex their market muscles.

Method of marketing and pricing

The influence of merchant activity is only felt in times of shortage. The major producers normally have a good deal of success in controlling prices by using an official producer price when contracting with consumers, often on a long-term basis When producers decide to or are forced to restrict sales due to increased demand, merchants are ready to trade at a premium to the official price. Other sources of material during these times would come from surplus consumers' stocks and other smaller producers.

Zinc

Grades available

The most commonly traded forms of zinc are 'electrolytic' with the purity of 99.95 per cent minimum, 'GOB' standing for good ordinary brand, which is 98.5 per cent minimum usually debased with lead, and special high grade with the purity greater than 99.99 per cent. All these types are traded in the form of ingots.

Production method

There are a number of different zinc-containing ores but the most commonly commercialized deposits are sulphide ores and the less common oxide ores.

Zinc ores often occur in combination with lead ores and with minor quantities of rarer metals, such as cadmium and indium. These other metals are separated and extracted during the refining process.

The sulphide ore is first roasted to remove sulphur. This sulphur is used to make sulphuric acid which can then be used to dissolve the zinc contained in the ore, to produce zinc sulphate. The zinc in turn is deposited from the solution onto cathodes in the electrolytic refining process. Zinc can also be refined by distillation, where the ore is roasted with carbon and the distillate collected. In both methods the refined metal is cast into slabs.

Major uses

One of zinc's most important uses is in galvanizing, the process of covering steel with a layer of zinc for protection from corrosion. This is commonly used to protect roofing sheets, girders, nails, ladders and buckets etc, and can be achieved by dipping the item in the molten metal or by electrolytic plating. Another major use is in brass, its alloy with copper.

Zinc and zinc alloys are also used extensively to produce die-casting alloys where they compete fiercely with aluminium. Zinc is heavier, but aluminium more costly. Zinc sheet has many uses, the most important being the cases for dry-cell batteries. Zinc oxide is used extensively in the rubber industry and in making white paint and pigments.

Zinc mine production 1977

metric tons x 1,000

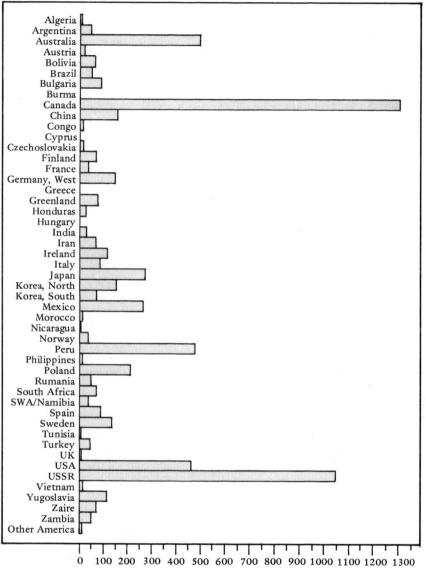

Total world production 6,593,300 metric tons

Main market features

Zinc is produced mainly in developed countries, notably Canada which is by far
the largest producer, although, as with a number of other metals, future supplies

are likely to come more from new deposits in the developing world. So far only Peru and Mexico are significant zinc producers in the third world. The Soviet Union is the world's second largest zinc producer and the largest producer in the communist bloc though North Korea and Poland are the major exporters to the Western World. A significant proportion of the world's zinc metal output, however, is concentrated in Western Europe, with smelters using zinc concentrates bought from mine producers. US zinc smelter production, however, has fallen sharply in the past decade, and the country is now far more reliant on imports. Strict anti-pollution regulations introduced by the government have made a number of US smelters uneconomic to operate, and during a period of low prices many of the old fashioned horizontal retort smelters were shut down and have not been replaced. The emphasis in new smelters all over the world is to switch to processes with less pollution. As a result modern smelters are tending to concentrate on zinc output, with by-products, rather than the older smelters producing both lead and zinc taken from the same ore.

An important feature of zinc is that, unlike copper, aluminium and lead, there is only a limited recovery of secondary, or scrap, zinc. This is because the main outlet is in galvanizing steel, requiring only a thin layer which is difficult to recover. However, the growth of the zinc die-casting market has created extra scope for increased scrap recovery, but this forms only a small part of the total market.

Known world reserves

125 million metric tons

Method of marketing and pricing

The bulk of the non-communist world's zinc mine and metal production is sold under direct supply contracts concluded on an annual basis between producers or smelters and consumers. The tonnage to be supplied is agreed, and the relevant published producer price quotation is used as a pricing basis. There are two main producer price quotations. One, in cents per pound, is used by US and Canadian producers and applies to the North American markets. The other is the so-called European producer price that is used for supply contracts in the rest of the world. Although known as the European price, it is in fact used by producers and smelters throughout the world as a basis for pricing ore concentrates as well as the refined metal.

The producer price for zinc, especially the European quotation, has come under increasing pressure in recent years. Producers and smelters have found it more difficult to cooperate amongst themselves. The mine producers are concerned with the overall return from selling the mine production in the form of ore concentrates, or from their smelters production of metal. But the independent smelters, unconnected with mining, are only concerned with achieving a profit between the cost of buying in the concentrates and selling the smelted metal. In weak market conditions, therefore, there is pressure by the smelters to reduce the cost of concentrates to maintain their margins, if they are forced to discount the selling price of the smelted metal. But producers are naturally opposed to this reduction in their overall return. The fragile relationship between producers and smelters in a number of countries was further undermined by the threat of anti-cartel action by the US government and the EEC Commission.

As a result the unofficial producers group was unable to take positive action to protect the European producer price, when a large surplus of supplies developed in 1977 as a result of a slump in demand, that lasted much longer than expected. The producer quotation became a 'nominal' quotation and the real prices being charged were reflected on the London Metal Exchange 'free' zinc market that has assumed a much greater importance. Since most zinc is sold under direct supply contracts, with little passing through the Exchange, prices remain, to a large extent, under the control of producers. In 1978 the producer price was reduced to more realistic levels, production cutbacks were implemented and producer control again asserted over the market. But the earlier virtual collapse of the European producer quotation demonstrated its inability to cope with a changing market situation adequately and brought more interest in the Metal Exchange free market.

Zinc prices

£ per metric ton. LME quotation.

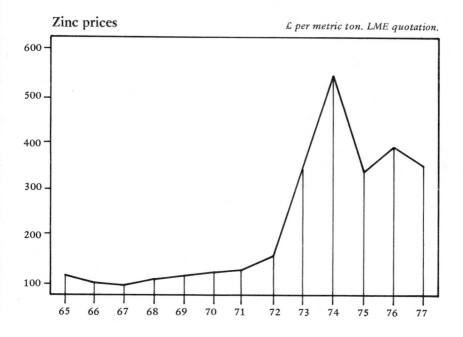

Zirconium

Zirconium ore production 1975

metric tons x 10,000

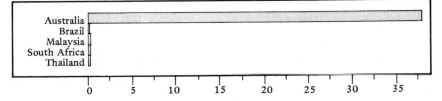

Total world production 408,351 metric tons

Grades available

The most common zirconium containing substances are its alloys some of which are known as zircaloy containing small quantities of tin or tin with iron, chrome and nickel. Zirconium metal is usually sold in the form of sponge and ingot.

Production method

Zirconium's ore, zircon, is roasted with coke in an electric arc furnace which produces zirconium carbonitride. This is then chlorinated to produce the chloride which is in turn reacted with magnesium metal to produce zirconium metal.

Major uses

Zircon, the ore, has a major use in its own right for making foundry moulding sands.

Several zirconium compunds have uses for making waxes, tanning agents, rust and water repellents, and deodorants.

The alloys have uses in the manufacture of pipes, pumps, valves and fittings in the chemical industry where its anti-corrosion properties are required.

Some pure metal is used in the construction of components for nuclear reactors and for making foil used in flash bulbs.

Main market features

Zirconium is produced by very few manufacturers. Its raw material is plentiful and very little of it is used to produce the metal. It is rather expensive to produce when one considers that another metal, magnesium, is required to make it. Its price has been rising steadily over the years in line with production costs. This all makes the market rather unexciting for metal merchants who confine their activity to trading in the ore or the scrap alloy.

No drastic upsurge in demand is foreseen but a major expansion in nuclear plant construction could cause a temporary lack of production capacity.

Method of marketing and pricing

The few manufacturers that there are sell directly to consumers. They also offer a range of alloys and in most cases semi-fabricated products as well. These producers determine what price they can charge and competition between them is not particularly fierce.

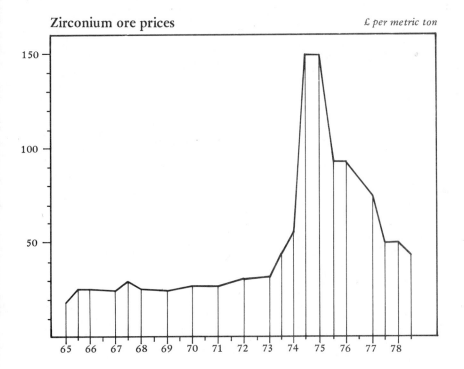

Zirconium ore prices   *£ per metric ton*

Appendices

Appendix 1:
Glossary of relevant terms

Actuals: Physical commodities, also commodities readily available. The commodity itself as opposed to a future contract

Agent: A merchant who acts on behalf of a producer usually on a commission basis

Anode: The electrode at which negatively charged ions are discharged during electrolysis

Arbitrage: Purchase of contract in one market while simultaneously selling the same amount in another to take advantage of price differentials

Backwardation: Market description of the situation when the spot or nearby prices are higher than those for future delivery months. Usually caused by delays in shipment thus creating shortages in available supplies. Opposite of contango

Basic price: Agreed price between buyer and seller of an option at which the option may be taken up. Also called the 'striking price'

Bear: Person expecting a decline in prices

Bear covering: Closing of short positions

Bid: The price which the buyer is willing to pay

Borrowing: Purchase of a nearby delivery date and simultaneous sale of a forward date. Used only in London Metal Exchange

Brine: Solution of any salt

Broker: Establishes contact between buyer and seller, for a fee. In the US, ring dealing members of futures markets are frequently called brokers

Bulk: Delivery of unpacked metal

Bull: Person expecting a rise in prices

Call: A period for trading. Conducted by a chairman to establish a price for a specific time. During a call, trading is confined to one delivery month

Call option: The option buyer/taker pays a premium and holds the right to decide at a later stage whether or not to buy at the price agreed at the time the premium was paid. The right may be exercised at any time from the point of purchase to the expiry of the option

Carrying: General term covering both *Borrowing* and *Lending*

Carrying Costs: Costs connected with warehouse storage, insurance, etc. On occasion includes interest and estimated changes in weight

Casting: Metal object made by pouring molten metal into a mould, which does not involve mechanical work such as rolling or forging

Catalyst: A chemical (metal) which assists a chemical reaction but remains chemically unchanged at the end

Cathode: The electrode at which positively charged (metallic) ions are discharged during electrolysis, hence the slab of metal deposited at the end of the electrolytic process

Certified stocks: Supplies rates as deliverable

C&F: Cost and freight

CIF: Cost, insurance and freight (included in price)

Clearing house: The organization that provides clearing facilities for some futures markets

Commission house: A company, which trades on behalf of clients for a commission. The Commission House only handles clients' business and does not trade on its own account

Concentrate: The ore of a metal after separation from other unwanted minerals

Conductivity: A measure of the ability of a metal to conduct an electric current

Contango: A situation where prices are higher in the forward delivery months than in the nearby delivery month. Opposite of backwardation. Normally in evidence when supplies are adequate or in surplus. The contango reflects either wholly or in part the costs of holding and financing

Contract: An agreement to buy or sell a specified amount of a particular commodity. It details the amount and grade of the product and the date on which the contract will mature and become deliverable, if not previously liquidated

Contract month: Month in which a given contract becomes deliverable, if not liquidated or traded out before the date specified

Custom smelter: A smelter which relies on concentrate purchased from independent mines instead of its own captive sources

Delivery basis: Specified locations to which the commodity in a futures contract may be physically delivered in order to terminate the contract

Delivery date: Or Prompt Date, on which the commodity must be delivered to fulfil the terms of the contract

Delivery month: Calendar month stipulated as month of delivery in a futures contract

Deposit: Sum of money required by the broker from his client, usually 10 per cent of the value of the contract, to justify opening of a futures position

Differentials: Premiums paid for grades better than the basic grade, or discounts allowed for grades below the basic grade

Double option: This is an option which gives the buyer or person taking the option the right either to buy from or sell to the seller of the option or the person who gives it, at the basis price

Ductility: Quality of a metal by which it may be beaten into sheet or drawn into a wire

Electrolysis: A process used for refining metals where the metal is deposited on a cathode from a solution or molten mass

Element: A substance which cannot be split into anything simpler in a chemical process

Fabricator: A company which makes semi-fabricated products from refined metal and on occasions from scrap

Fixation: Fixing price in the future, and used in commodity call purchases and call sale trades

Flotation: Process by which ore is separated from earth and rubble, during which the ore is introduced to a bath of liquid which is agitated so that the metallic ore either rises to the top or sinks to the bottom

Flux: A substance added to a metal in order to make it easier to melt

FOB: Free on Board

Force Majeure: This is a clause in a supply contract which permits either party not to fulfil the contractual commitments due to events beyond their control. These events may range from strikes to export delays in producing countries

Forward shipment: Contract covering actual commodity shipments at a specified date in the future

Futures contract: Contract which requires the delivery of a commodity in a specified future month, if not liquidated before the contract matures

Grades: Standards set for judging the quality of a commodity

Hedge: A temporary futures market sale which is made against a spot purchase, or alternatively a temporary futures market purchase made against a spot sale. The purpose is to reduce risk from price fluctuations on the physical transaction until the reverse futures market operation cancels the hedge, or liquidates the original operation

Inorganic: Compounds other than those of carbon, but including the oxides of carbon and carbonates

Integrated producer: A producer who owns mines, smelters and refineries and also, in some instances, fabricating plants

Kerb trading: Unofficial trading when the market has closed. The term 'kerb' dates from the time when dealers continued trading on the kerb outside the exchanges after these had closed

Last trading day: Final day for trading a particular delivery. Positions which have not been closed by the last trading day must be fulfilled by making, or taking, delivery of the physical commodity or metal

Leaching: The process by which metals are extracted from a low grade ore or waste product, by the direct action of a caustic solution or an acid, producing a metallic salt solution

Lending: Sale of a nearby delivery date coupled with the simultaneous purchase of a more distant date (LME term)

Liquidation: Sale of long contract to offset previous purchase. Operation which cancels an earlier position

Long: An open purchased futures contract. Buying forward on the market

Lot: Minimum contract unit in a hedge or futures market

Margin: This is the amount deposited as a guarantee for the fluctuations on a futures purchase or sale. If the contract fluctuates against the holder of the contract, he is required to provide for the difference between his contract price and the current market price by paying 'variation margin' differences. Thus the original margin continues to guarantee fully the performance of the contract at any market price level

Margin call: A commodity broker's request to a client for additional funds to secure the original deposits

Master alloy: An alloy made with two metals x and y, which is then added to a larger quantity of metal x so that metal y is diluted to the required proportion

Nearby delivery: The nearest active month of delivery on a futures market

Offer: The seller's price for the commodity offered

Open contracts: Contracts bought or sold and not offset by an opposite trade

Open outcry: Trading conducted by calling out bids and offers across a ring or pit and having them accepted

Open position: A forward market position which has not been closed out

Option: The holder of the option has the right to buy from or sell to the granter of the option a specified quantity of the commodity at an agreed price. The cost of buying the option is known as the premium

Pegged price: The price at which a commodity has been fixed by agreement

Plating: The deposition of a layer of a metal on an object forming the cathode during electrolysis

Position trader: Someone who takes long or short positions in futures markets in consequence of an opinion that prices are about to advance or decline

Premium: The amount by which a cash commodity price sells over a futures price or another cash commodity price. The excess of one futures contract price over another

Put option: This option gives the buyer (or 'taker') of the option — in exchange for the premium which he pays — the right to decide at a later date whether or not to sell to the seller (or 'granter') at the price ('basic' or 'striking' price) agreed at the time the premium was paid. The right may be exercised at any point from the purchase of the option up to the declaration date (the date upon which the option expires)

Reduction: A process involving the removal of oxygen or a group which forms negative ions

Ring: Space on a trading floor where futures are traded. Also known as a pit in the US

Roast: Heating ore or concentrate in air or inert atmosphere to extract unwanted element, usually sulphur

Salt: A compound containing a metal and a non-metal

Scavenger: Substance added to molten metal which has the effect of ridding the melt of an unwanted element

Semi-fabrication: The production of metal in the form of wire, sheet, rod etc

Short: The sale of a commodity not owned by the seller

Slime: Substance containing metal other than that deposited at the cathode during electrolysis and which is found as a sludge in the bottom of the electrolytic cell

Smelt: Extraction of crude metal from the ore by melting, prior to refining

Sponge: Lumpy form of metal with sponge-like appearance produced by casting molten metal into water

Spot: Term denoting immediate delivery for cash, as distinct from future delivery

Spot month: The first month in which delivery can take place and for which a quotation is made on the futures market

Spot price: The commodity cash sale price, as opposed to a futures price

Spread: An order to purchase one contract month and sell another month in the same commodity, usually done on the same exchange

Squeeze: Pressure on a delivery date which results in the price of that date becoming firmer in relation to other dates

Straddle: The simultaneous buying and selling of the same commodity on the same market. This is designed to take advantage of differences between two options. Example – the sale of a September option and the simultaneous purchase of a January option made in the expectation that a later simultaneous purchase of the September and sale of the January options will produce a profit. See Arbitrage

Sublimation: Heating an unrefined metal until it turns into its gaseous form. This is then condensed into the refined metal

Switch: To advance or postpone the original contract to a different month

Switching: Exchanging a commodity in one warehouse for a commodity in another

Terminal market: Usually synonymous with commodity exchange or futures market, especially in the United Kingdom

Unwrought: Metal in a cast condition, unworked by mechanical means

Warehouse receipt: A receipt for a commodity given by a licensed or authorized warehouseman and issued as tender on futures contracts

Warrant or Warehouse receipt: A receipt of physical deposit which gives title to the physical commodity

Appendix 2:

Average monthly dollar/sterling rate based on the market price at 11.30 a.m. each dealing day

	January	February	March	April	May	June	July	August	September	October	November	December	Average
1958	2.8000	2.8000	2.8000	2.8000	2.8000	2.7999	2.7979	2.7978	2.7981	2.7980	2.7995	2.7994	2.7992
1959	2.8047	2.8077	2.8103	2.8148	2.8127	2.8109	2.8101	2.8080	2.8020	2.8050	2.8012	2.7970	2.8070
1960	2.7985	2.8017	2.8041	2.8078	2.8046	2.8010	2.8065	2.8084	2.8107	2.8096	2.8119	2.8057	2.8059
1961	2.8047	2.7986	2.7956	2.7965	2.7922	2.7888	2.7854	2.8006	2.8096	2.8136	2.8130	2.8078	2.8005
1962	2.8089	2.8126	2.8132	2.8122	2.8105	2.8068	2.8050	2.8023	2.7995	2.7999	2.8001	2.8019	2.8061
1963	2.8031	2.8018	2.7990	2.7990	2.7980	2.7986	2.7993	2.7981	2.7961	2.7964	2.7964	2.7951	2.7984
1964	2.7967	2.7955	2.7966	2.7974	2.7979	2.7926	2.7884	2.7852	2.7819	2.7820	2.7833	2.7887	2.7905
1965	2.7895	2.7932	2.7904	2.7938	2.7953	2.7904	2.7892	2.7891	2.7945	2.8014	2.8020	2.8003	2.7941
1966	2.8025	2.8010	2.7938	2.7918	2.7908	2.7884	2.7873	2.7875	2.7877	2.7901	2.7896	2.7884	2.7916
1967	2.7892	2.7924	2.7943	2.7974	2.7952	2.7899	2.7860	2.7838	2.7829	2.7817	2.6430	2.4062	2.7452
1968	2.4087	2.4094	2.3992	2.4015	2.3891	2.3845	2.3892	2.3912	2.3869	2.3895	2.3874	2.3838	2.3934
1969	2.3866	2.3910	2.3913	2.3928	2.3862	2.3896	2.3902	2.3855	2.3837	2.3896	2.3962	2.3967	2.3900
1970	2.3999	2.4042	2.4056	2.4059	2.4033	2.3976	2.3903	2.3877	2.3845	2.3870	2.3899	2.3903	2.3955
1971	2.4047	2.4170	2.4182	2.4175	2.4183	2.4179	2.4182	2.4342	2.4709	2.4915	2.4937	2.5267	2.4441
1972	2.5703	2.6030	2.6178	2.6093	2.6117	2.5814	2.4415	2.4502	2.4425	2.3968	2.3518	2.3456	2.5018
1973	2.3558	2.4418	2.4778	2.4825	2.5174	2.5743	2.5402	2.4742	2.4163	2.4260	2.3890	2.3171	2.4510
1974	2.2263	2.2738	2.3369	2.3906	2.4117	2.3876	2.3869	2.3445	2.3146	2.3310	2.3271	2.3271	2.3382
1975	2.3603	2.3917	2.4172	2.3680	2.3185	2.2807	2.1833	2.1133	2.0852	2.0548	2.0499	2.0208	2.2203
1976	2.0276	2.0257	1.9455	1.8476	1.8156	1.7710	1.7863	1.7819	1.7309	1.6401	1.6370	1.6787	1.8073
1977	1.7127	1.7100	1.7172	1.7195	1.7188	1.7193	1.7221	1.7403	1.7436	1.7703	1.8203	1.8550	1.7458
1978	1.9338	1.9405	1.9103	1.8507	1.8175	1.8366	1.8940	1.9419	1.9593	—	—	—	—

Appendix 3:

World time zones

Standard Times at noon Greenwich Mean Time

British Summer Time, which is one hour ahead of GMT, is observed from 02.00 hours on the third Sunday in March until 02.00 on the fourth Sunday in October. USA Daylight Saving Time, which is one hour ahead of local standard time, is observed in all states except Arizona, Hawaii and Michigan from 02.00 on the last Sunday in April until 02.00 on the last Sunday in October. In France, in 1977, Summer Time, which is also one hour ahead of local standard time, is observed from 24.00 hours on 3 April to 24.00 on 25 September.

Accra	12.00	Damascus	14.00	Ottawa	07.00
Adelaide	21.30	Darwin	21.30	Panama	07.00
Algiers	13.00	Delhi	17.30	Paris	13.00
Amman	14.00	Djakarta	20.00	Peking	20.00
Amsterdam	13.00	Dublin	12.00	Perth	20.00
Ankara	14.00	Gibraltar	13.00	Prague	13.00
Athens	14.00	Helsinki	14.00	Quebec	07.00
Auckland	24.00	Hobart	22.00	Rangoon	18.30
Baghdad	15.00	Hong Kong	20.00	Rawalpindi	17.00
Bangkok	19.00	Istanbul	14.00	Reykjavik	12.00
Beirut	14.00	Jerusalem	14.00	Rio de Janeiro	09.00
Belgrade	13.00	Karachi	17.00	Rome	13.00
Berlin	13.00	Kuala Lumpur	20.00	San Francisco	04.00
Berne	13.00	Lagos	13.00	Santiago	08.00
Bombay	17.30	Leningrad	15.00	Sofia	14.00
Bonn	13.00	Lima	07.00	Singapore	19.30
Brisbane	22.00	Lisbon	13.00	Stockholm	13.00
Brussels	13.00	Luxembourg	13.00	Sydney	22.00
Bucharest	14.00	Madras	17.30	Tehran	15.30
Budapest	13.00	Madrid	13.00	Tokyo	21.00
Buenos Aires	09.00	Melbourne	22.00	Toronto	07.00
Cairo	14.00	Mexico City	06.00	Tunis	13.00
Calcutta	17.30	Montevideo	08.30	Vancouver	04.00
Canberra	22.00	Moscow	15.00	Vienna	13.00
Cape Town	14.00	Nairobi	15.00	Warsaw	13.00
Caracas	08.00	New York	07.00	Washington	07.00
Chicago	06.00	Nicosia	14.00	Wellington	24.00
Copenhagen	13.00	Oslo	13.00	Winnipeg	06.00

Note to World Time Zones map
The earth turns one complete revolution in 24 hours. The surface of the Earth is divided into 24 Time Zones, each of 15° longitude or 1 hour of time. In 24 hours it turns through 360°. The times shown are the standard times on land and sea when it is 12.00 hours on the Greenwich Meridian.

World Time Zones

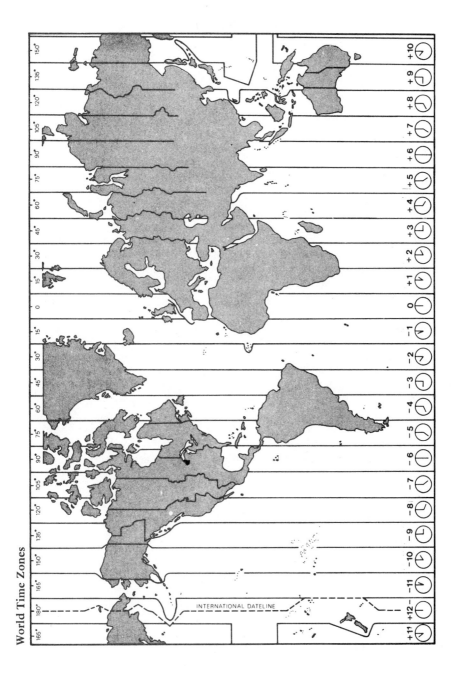

INTERNATIONAL DATELINE

171

Appendix 4:
London Metal Exchange contract rules and arbitration

Contract rules

Rule A
Members of the London Metal Exchange, in their dealings with other Members, shall be responsible to and entitled to claim against one another, and one another only, for the fulfilment of every Contract for Metals.

Rule B
In these Rules the expression 'Members of the London Metal Exchange' includes Firms and Companies who, although not themselves Subscribers to the Exchange, are represented and deal thereon by and through 'Representative Subscribers' to the Exchange acting as the representatives or Agents of such Firms or Companies.

Rule C
If any Member of the Metal Exchange fails to meet his engagements to another Member, whether by failing to provide on the due date documents (i.e. Bills of Lading, Warrants or Delivery Orders according to the metals dealt in) to meet sales made or money to pay for metals bought, or by making default in fulfilling any other obligation arising out of dealings made subject to the Rules and Regulations of the London Metal Exchange, notice of the default shall be given at once in writing to the Committee of the Exchange and the Committee shall immediately fix and publish a settlement price or prices as at the date of such communication to them for all contracts which the defaulter may have open under these Rules, whether with Members or with parties who are not Members. All such contracts shall forthwith be closed and balanced, by selling to or buying from the defaulting Member such metals as he may have contracted to deliver or take, at the settlement prices fixed for this purpose by the Committee, and any difference arising whether from or to the party in default shall become payable forthwith notwithstanding that the prompt day or other day originally stipulated for the settlement of the transaction may not have arrived. In fixing settlement prices under this Rule the Committee may in their discretion take into consideration the extent and nature of the transactions which the defaulting Member has open and any other circumstance which they may consider should affect their decision. In any case where the Committee shall be of opinion that the default is not due to the insolvency of the defaulter the Committee shall by resolution negative the application of this rule. Any claim arising out of a default not due to insolvency shall be settled by arbitration in the usual manner. This rule shall

apply to cases in which at or after the decease of a Member the engagements entered into by him are not duly met.

Rule D

In any Contract made subject to the Rules and Regulations of the London Metal Exchange between a Member and a non-Member, in the event of the Non-Member failing to meet his engagement arising out of any such contract whether by failing to provide on the due date documents to meet sales or money to take up documents (as the case may be) or otherwise howsoever or of his failing to supply or maintain such margin (if any) for which the Member is entitled to call and has called, or in the event of the Non-Member's suspending payment or becoming bankrupt or committing any act of bankruptcy or (being a Company) in the event of its going into liquidation whether voluntary or otherwise, the Member shall have the right to close all or any such Contracts outstanding between them by selling out or buying in against the Non-Member (as the case may be) and any differences arising therefrom shall be payable forthwith notwithstanding that the prompt day or other day originally stipulated for settlement may not have arrived.

Rule E

Payments for Warrants or other documents (when deliverable under the Contracts) unless otherwise stipulated on the contract, shall be made by cash in London by cheque on a London *clearing* bank, either mode in Seller's option. The documents shall be tendered in London against the cash or cheque, as the case may be, and not later than 2.30 p.m. on the prompt or settling day.

Rule F

Contracts wherein Buyer or Seller (as the case may be) has the option to uplift or to deliver, prior to the prompt or settlement date by giving previous notice of his intention, shall have the notice reckoned by market days; such notices, unless otherwise stipulated at time of purchase or sale, shall be as follows: On a Contract with the option to uplift or to deliver during one calendar month or less, one day's notice shall be given; on a Contract with the option beyond one and up to two calendar months two days' notice shall be given; and on a Contract with the option beyond two and up to three calendar months three days' notice shall be given previous to the date on which delivery is required, or will be made. In the case of Silver on a Contract with the option beyond three and up to seven calendar months seven days' notice shall be given previous to the with the option beyond four and up to five calendar months five days' notice shall be given; on a Contract with the option beyond five and up to six calendar months six days' notice shall be given; on a Contract with the option beyond six and up to seven calendar months seven days' notice shall be given previous to the date on which delivery is required, or will be made. Notice shall be given for the whole quantity stated in the contract and shall be tendered in writing and delivered at the office of the seller of the option not later than noon on the day of notice. Rent shall only be allowed to Buyer to actual day of settlement; and

173

there shall not be any allowance of interest for a payment made prior to the prompt date.

Rule G

Prompt or settlement dates falling on Saturday, Sunday, or a Bank Holiday which days are not market days, shall be settled as follows. Prompts falling on Saturday shall be settled on the Friday previous; but should the preceding Friday be a Bank Holiday the prompt shall be extended to the Monday following; should both the Friday preceding and the Monday following be Bank Holidays, the prompt shall be settled on the Thursday previous. Prompts falling on Sunday should be extended to the Monday following, but should that Monday be a Bank Holiday the prompt shall be extended to the Tuesday following; should both the following Monday and Tuesday be Bank Holidays, the prompt shall then be extended to the Wednesday following. Prompts falling on a Bank Holiday shall be extended to the day following; and if the Bank Holiday fall on Friday the prompts shall be extended to the Monday following; but should the Friday be Good Friday, prompts falling on that day shall be settled on the Thursday previous. If Christmas Day falls on Monday, prompts falling on that day shall be extended to the Wednesday following, but if Christmas Day falls on Tuesday, Wednesday, Thursday, or Friday, prompts falling on that day shall be settled on the day previous.

Rule H

The establishment, or attempted establishment of a 'corner', or participation directly or indirectly in either, being detrimental to the interest of the Exchange, the Committee shall, if in their opinion a 'corner' has been or is in the course of being established, have power to investigate the matter and to take whatever action it considers proper to restore equilibrium between supply and demand. Any member or members may be required to give such information as is in his or their possession relative to the matter under investigation.

Rule J (OPTIONS)

On the day on which notice is due, the holder of the option shall, except in cases to which Rule C applies, declare in writing before 12 o'clock noon whether he exercises or abandons the option, and if he fails to make such declaration the option shall be considered as abandoned. Options (subject to Rule F above) may be declared for less than the total optional quantity in quantities of 25 tonnes for Copper-Electrolytic Wirebars, H.C.F.R. Wirebars, Cathodes or Fire Refined, 5 tonnes for Standard Tin, High Grade Tin, 25 tonnes for Standard Lead, 25 tonnes for Standard Zinc and 10,000 ounces for Silver or multiples thereof, only one declaration against each contract being allowed. In cases to which Rule C applies the prices fixed by the Committee, at which outstanding contracts are to be closed, shall equally apply to all option contracts; and all options shall be automatically determined, and be deemed to have been either exercised or abandoned according as the prices may be in favour of or against the defaulter and whether the defaulter be the Seller or the Buyer of an option, and the option money shall be brought into account. In contracts with optional prompts, the price which shall be taken as the basis of settlement shall be the settlement price fixed by the Committee under Rule C for the prompt most favourable to the holder of the option.

Rule K (CLEARING)
All contracts made between Members of the London Metal Exchange who are entitled to deal in the Ring, either for Copper-Electrolytic Wirebars, H.C.F.R. Wirebars, Cathodes or Fire Refined, Standard Tin, High Grade Tin, Standard Lead, Standard Zinc or Silver, shall be settled through the Clearing, except when a Member insists on his right to receive cash instead of cheque from the Member to whom he has sold, in which case the Seller shall give notice to his Buyer before noon on the market day preceding the settling day, and such transactions shall then be exempted from settlement through the Clearing. The Rules governing the Clearing of all contracts shall be those in existence at the time fixed for the fulfilment of the contract. Copies of such rules may be obtained from the Secretary of the Exchange.

Rule L
In the case of strikes, lock-outs, or other unforeseen contingencies in London, or other authorised port or point of delivery, which prevent or delay the discharge and/or warehousing of Copper-Electrolytic Wirebars, H.C.F.R. Wirebars, Cathodes or Fire Refined, Standard Tin, High Grade Tin, Standard Lead, Standard Zinc and/or Silver, the Seller may be allowed to postpone delivery if he can prove to the satisfaction of the Committee (of which proof the Committee shall be the sole judge) that he does not hold available metal in warehouse or vault with which to fulfil his contracts and that he has metal of the requisite quality which has arrived in London or any other authorised port or point of delivery at least ten days prior to the earliest prompt for which relief is asked, or has metal of the requisite quality in his works, but the delivery, discharge and/or warehousing of which is prevented or delayed as aforesaid. He must also deposit with the Secretary of the Exchange such sums as the Committee may require but not exceeding £5 per tonne in the case of Copper, Lead and Zinc, £10 per tonne in the case of Tin and £5 per thousand ounces in the case of Silver. No interest will be allowed on deposits, which will be returned after delivery of Warrants. Should his application be passed by the Committee, he shall deposit documents or other proof to the satisfaction of the Committee with the Secretary of the Exchange, who shall issue Certificates for Copper, Lead and Zinc in quantities of 25 tonnes, Certificates for Tin in quantities of 5 tonnes and Certificates for Silver in quantities of 10,000 ounces. The Seller shall deliver these Certificates to his Buyer. The Certificates will then constitute a good delivery on the Clearing within the period stated thereon and differences must be settled on the prompt day. The holder of a Certificate must present it to the firm named thereon not later than 2.30 p.m. on the day following that on which he receives notice in writing from his Seller that the Warrant for the actual Copper, Tin, Lead, Zinc or Silver is ready. He must take up the Warrant against payment at the settlement

price fixed on the preceding market day, receiving or paying any difference between this and the price mentioned on the Certificate. In the event of the price on the Certificate being above or below the settlement price operative on the day of delivery, the receiver shall pay or be paid the amount of any difference. No other payments shall pass except against delivery of the actual Warrant. In case of any dispute, the Committee's ruling to be final. A fee of £5 to be paid by the Applicants for each Certificate issued.

Arbitration

Rule 1
All disputes arising out of or in relation to contracts subject to the Rules and Regulations of the London Metal Exchange shall be referred to arbitration as hereinafter provided. The Executive Secretary of the Committee of the London Metal Exchange (hereinafter referred to as 'the Secretary') shall be notified of such disputes in writing and the party first notifying the difference shall at the time of such notification deposit with the Metal Market & Exchange Co. Ltd., the sum of £100. All such disputes shall be referred to two arbitrators, one to be appointed by each party to the difference from the Arbitration Panel of the London Metal Exchange, such arbitrators having power to appoint a third arbitrator from the Panel and having all the powers conferred on arbitrators by the Arbitration Act 1950 or any statutory modifications thereof for the time being in force. The Secretary shall be notified in writing by each party of the appointment of the arbitrators. The arbitration and any Appeal made pursuant to Rule 8 of these Rules from the Award of the Arbitrators to the Committee shall take place at the London Metal Exchange (unless mutually agreed by the Arbitrators and the parties to the dispute that the venue should be elsewhere in England or Wales) and English procedure and law shall be applied thereto.

Rule 2
Persons eligible for appointment to the Arbitration Panel shall be members of the Exchange, their partners or co-directors (as the case may be) or members of their staff. Appointment to and removal from the Panel shall be made, at their sole discretion, by the Committee of the London Metal Exchange who will also be responsible for maintaining a panel of sufficient size.

Rule 3
In the event of either party to the difference (a) failing to appoint an arbitrator, or (b) failing to give notice in writing or by cable of such appointment to reach the other party within 14 days after receiving written or cabled notice from such other party of the appointment of an arbitrator (any notice by either party being given to the other either by cable or by registered post addressed to the usual place of business of such other party), or (c) in the case of death, refusal to act, or incapacity of an arbitrator, then, upon written or cabled request of either party an arbitrator shall be appointed from the said Arbitration Panel by the Committee of the London Metal Exchange.

Rule 4

In case the two arbitrators appointed as aforesaid, whether originally or by way of substitution, shall not within three calendar months after the appointment of the arbitrator last appointed deliver their Award in writing, or choose a third arbitrator, then the said Committee on the written request of either party shall appoint a third arbitrator selected from the said Arbitration Panel to act with the two aforesaid arbitrators.

Rule 5

The Award in writing of the arbitrators or any two of them shall be made and delivered in triplicate to the Secretary within a period of three calendar months from the date of the acceptance of the appointment by the arbitrator last appointed.

Rule 6

Every Award made pursuant to any provision of this Rule shall be conclusive and binding on the parties to the arbitration, subject to appeal as hereinafter mentioned.

Rule 7

The procedure upon an arbitration shall be as follows:

(a) Within a period of 21 days after the appointment of the second of the two arbitrators so appointed, each party shall deliver to the arbitrators and to each other a statement of case in writing with the originals, or copies, or any documents referred to therein. All such documents to be in the English language or accompanied by certified translations into English.

(b) If either party shall make default in delivering such statements and documents (due consideration being given to time occupied by mails) the arbitrators shall proceed with the case on the statement before them, provided always that, in the sole discretion of the arbitrators, an extension of time may be allowed for the delivery of such statements and documents.

(c) The arbitrators shall appoint a day for a hearing within 28 days, or such further time as the arbitrators shall in their sole discretion allow, after the expiry of the 21 days in accordance with Rule 7(a), and shall give due notice in writing thereof to the parties, who may, and if required by the arbitrators shall, attend and shall submit to examination by the arbitrators and produce such books and documents as the arbitrators may require. Each party shall be entitled to produce verbal evidence before the arbitrators.

(d) Neither Counsel, nor Solicitor shall be briefed to appear for either party without the consent of the arbitrators.

(e) The arbitrators may engage legal or other assistance.

177

(f) The arbitrators may adjourn the hearing from time to time, giving due notice in writing to the parties of the resumed hearing, and the arbitrators may, if they think fit, proceed with such a resumed hearing in the absence of either party or of both parties.

(g) Where any change takes place in the constitution of the tribunal of arbitrators, either by substitution or otherwise, the new tribunal shall appoint a day for the hearing which shall be not later than 28 days, nor earlier than 7 days, after the change. Each party, if desiring to do so, may submit an Amended Statement of Case, with a copy to the other party, which must reach the new tribunal within seven days of its appointment.

(h) In the event of a third arbitrator being appointed, the provisions contained in Section 9 Sub-Section 1 of the Arbitration Act 1950 shall not apply to any reference.

(i) The cost of the arbitration shall be at the sole discretion of the arbitrators. The arbitrators shall fix the amount of their remuneration. The Award shall state separately the amount of such costs and remuneration and by whom they shall be paid and whether the whole or any part of the deposit referred to in Rule 1 of these Rules shall be returned to the party lodging the same or be forfeited. In the event of either or both parties having been granted permission by the arbitrators to be legally represented at the hearing, the arbitrators may take into consideration any legal costs which have been incurred.

(j) The Award shall be deposited with the Secretary who shall forthwith give notice of receipt thereof in writing to both parties, and a copy of such Award shall be delivered to both parties on payment by either party of the costs specified in the Award, which payment shall not affect any provision of the Award.

(k) In the event that after the deposit referred to in Rule 1 of these Rules has been made the parties to the arbitration shall (i) settle their differences; (ii) fail to proceed as directed by the arbitrators under sub-clause (c) of this Rule; (iii) fail to take up the Award within 28 clear days of notification being given under sub-clause (j) of this Rule, such deposit shall be forfeited.

(l) At the time of issuing their Award, all statements and all documents lodged with the arbitrators shall be delivered by them to the Secretary, by whom they shall be retained until the expiration of the time for giving notice of appeal, as hereafter mentioned, after which the Secretary shall, unless there shall be such appeal, return them to the parties concerned.

Rule 8
Either party shall have the right to appeal against the Award to the Committee of the London Metal Exchange.

Rule 9
The method of appeal against the Award shall be as follows:

(a) The party making the appeal shall (i) within 21 days of the date of the Award give notice in writing of such appeal to the Secretary, and to the other party and shall at the same time state the grounds for appeal. (ii) Deposit with the Secretary the sum of £200, and in addition the sum, if any, which shall be payable under the Award by the Appellant.

(b) Upon the receipt of such Notice of Appeal the Committee shall within 4 weeks nominate not less than five members (hereinafter called 'the Appeal Committee') to hear the Appeal. Members of the Appeal Committee shall be members of the Committee of the London Metal Exchange and/or members of the Board of the Metal Market & Exchange Co. Ltd.

(c) The procedure on appeal shall as far as possible be similar to that above provided for the original hearing, except that all statements and documents delivered to the Secretary under Rule 7(l) shall be laid before the Appeal Committee, who may, however, require such further statement or statements or other information or documents from either or both of the parties as the Appeal Committee may think necessary. The provision of Rule 7(k) shall apply in like manner to the deposit referred to in sub-paragraph (a)(ii) of this Rule as the deposit in connection with the original hearing.

(d) The decision in writing of the majority of the Appeal Committee (which latter shall not at any time number less than five) shall be final and binding on all parties, and the Appeal Committee shall also decide whether the whole or any part of the said deposit of £200 shall be returned to the Appellant or be forfeited.

(e) The Appeal Committee shall have the same discretion regarding costs as is given to the arbitrators under Rule 7(i) and shall fix the amount of their remuneration and direct by whom it shall be paid.

(f) All statements and all documents lodged with the Appeal Committee shall together with the Award, be deposited by them with the Secretary by whom they shall be retained until the costs and fees specified in the Awards have been paid by either party. On payment, which shall not affect any provision of the Award, a copy of the Award shall be delivered to both parties and all documents returned to the parties concerned.

Appendix 5:
World currencies

Country	Basic or principal unit of currency
Afghanistan	Afghani
Albania	Lekë
Algeria	Algerian Dinar
Angola	Kwanza
Antigua	East Caribbean Dollar (EC$)
Argentina	Argentinian Peso
Australia	Australian Dollar
Austria	Schilling
Bahamas, The	Bahamian Dollar (BS)
Bahrain	Bahrain Dinar
Bangladesh	Taka
Barbados	East Caribbean Dollar (EC$)
Belgium	Belgian Franc
Belize	Belizean Dollar
Benin	Franc CFA
Bermuda	Bermudan Dollar (Canadian, UK and US currencies also in use)
Bhutan	Tikchung
Bolivia	Bolivian Peso
Botswana	Pula
Brazil	Cruzeiro
Brunei	Brunei Dollar
Bulgaria	Lev
Burma	Kyat
Burundi	Burundi Franc
Cameroun Republic	Franc de la Communauté Financière Africaine (Franc CFA)
Canada	Canadian Dollar
Central African Empire	Franc CFA
Central American Common Market	Central American Peso
Chad	Franc CFA
Chile	Chilean Escudo

Country	Basic or principal unit of currency
China, People's Republic of	Renminbi Yuan
Colombia	Colombian Peso
Congo (Brazzaville)	Franc CFA
Costa Rica	Colon
Cuba	Cuban Peso
Cyprus	Cyprus Pound
Czechoslovakia	Koruna
Dahomey	Franc CFA
Denmark	Danish Krone
Dominica	East Caribbean Dollar (EC$)
Dominican Republic	Dominican Republic Peso
Ecuador	Sucre
Egypt	Egyptian Pound (£E)
El Salvador	El Salvador Colon
Ethiopia	Ethiopian Birr
Fiji	Fijian Dollar $F)
Finland	Markka
France	French Franc
French Guinea	French Franc
Gabon	Franc CFA
Gambia	Dalasi
German Democratic Republic	Ostmark or DDR-Mark
Germany, Federal Republic	Deutsche Mark (DM)
Ghana	Cedi
Gibraltar	Gibraltar Pound
Greece	Drachma
Grenada	East Caribbean Dollar (EC$)
Guadeloupe	French Franc
Guatemala	Quetzal
Guinea-Bissau	Escudo
Guinea, Equatorial	Guinean Peseta
Guinea, Republic of	Sili
Guyana	Guyanese Dollar $G)
Haiti	Gourde
Honduras	Lempira (L)
Hong Kong	Hong Kong Dollar (HK$)
Hungary	Forint
Iceland	Icelandic Krona
India	Indian Rupee
Indonesia	Rupiah (Rp)
Iran	Iranian Rial
Iraq	Iraqi Dinar

181

Country	Basic or principal unit of currency
Ireland	Irish Pound
Israel	Israeli Pound (I£)
Italy	Lira
Ivory Coast	Franc CFA
Jamaica	Jamaican Dollar (J$)
Japan	Yen
Jordan	Jordanian Dinar (JD)
Kenya	Kenya Shilling (Ks)
Khmer Republic	Riel
Korea, Democratic People's Republic of (North Korea)	Won
Korea, Republic of (South Korea)	Won
Kuwait	Kuwait Dinar (KD)
Laos	Kip Pot Poi
Lebanon	Lebanese Pound (£L)
Lesotho	South African Rand
Liberia	Liberian Dollar (also US Dollar)
Libya	Libyan Dinar (LD)
Liechtenstein	Franken or Swiss Franc
Luxemburg	Luxemburg Franc
Macao	Pataca
Malagasy Republic	Franc Malgache (FMG)
Malawi	Kwacha (K)
Malaysia	Ringgit
Mali	Mali Franc
Malta	Maltese Pound
Martinique	Local Franc
Mauritania	Ouguiya
Mauritius	Mauritian Rupee
Mexico	Mexican Peso
Monaco	French Franc
Mongolia	Tugrik
Montserrat	East Caribbean Dollar (EC$)
Morocco	Dirham
Mozambique	Escudo
Namibia (South West Africa)	South African Rand
Nepal	Nepali Rupee (NR)
Netherlands	Netherlands Gulden or Guilder or Florin
Netherlands Antilles	Netherlands Antilles Florin (NAFI) or Guilder
New Zealand	New Zealand Dollar ($NZ)

Country	Basic or principal unit of currency
Nicaragua	Cordoba
Niger	Franc CFA
Nigeria	Naira
Norway	Norwegian Krone
Oman	Ryal Omani
Pakistan	Pakistan Rupee
Panama	Balboa
Papua New Guinea	Kina
Paraguay	Guarani
Peru	Gold Sol
Philippines	Philippine Peso
Pitcairn Islands	Australian Dollar
Poland	Zloty
Portugal	Portuguese Escudo
Puerto Rico	US Dollar ($)
Qatar	Qatar/Dubai Ryal
Réunion	French Franc
Rhodesia	Rhodesian Dollar (R$)
Romania	Leu
Rwanda	Rwanda Franc
St Kitts	East Caribbean Dollar (EC$)
St Lucia	East Caribbean Dollar (EC$)
St Vincent	East Caribbean Dollar (EC$)
Samoa (Western)	Tala
Saudi Arabia	Saudi Arabian Ryal
Senegal	Franc CFA
Seychelles	Rupee
Sierra Leone	Leone
Sikkim	Indian Rupee
Singapore	Singapore Dollar (S$)
Somalia	Somali Shilling
South Africa, Republic of	Rand (R)
South West Africa (Namibia)	South African Rand
Spain	Spanish Peseta
Sri Lanka	Sri Lanka Rupee
Sudan	Sudanese Pound
Surinam	Surinam Guilder (also Netherlands currency)
Swaziland	Lilangeni
Sweden	Swedish Krona
Switzerland	Franken or Swiss Franc
Syria	Syrian Pound (£S)

Country	Basic or principal unit of currency
Taiwan	New Taiwan Dollar
Tanzania	Tanzanian Shilling
Thailand	Baht
Togo	Franc CFA
Tonga	Pa'anga
Trinidad and Tobago	Trinidad and Tobago Dollar
Tunisia	Tunisian Dinar
Turkey	Turkish Lira (TL)
Uganda	Ugandan Shilling
United Arab Emirates	Dirham
United Kingdom (UK)	Pound Sterling (£)
United States of America (USA)	US Dollar ($)
Upper Volta	Franc CFA
Uruguay	Uruguayan Peso
USSR	Rouble
Venezuela	Bolivar
Vietnam	Dong
Western Samoa	Tala
West Indies Associated States	East Caribbean Dollar (EC$)
Yemen Arab Republic	Yemeni Ryal (also Indian Rupee)
Yemen, People's Democratic Republic of	South Yemen Dinar
Yugoslavia	Yugoslav Dinar
Zaire	Zaire
Zambia	Kwacha (K)

Appendix 6:
Conversion table

Mass	Metric ton	Long ton	Short ton	Pounds	Kilograms	Ounce troy	Grains	Pennyweight	Pikul
Metric ton		0.984	1.102	2204.5	1000				16.535
Long ton	1.016		1.120	2240	1016				
Short ton	0.907	0.893		2000	907				15.000
Pound	0.0004	0.0004	0.0005		0.4536	14.583	7000		
Kilogram	0.001			2.2045		32.150			
Ounce troy							480	20	
Grain						480			
Pennyweight							24		
Pikul	0.0605		0.067	133.33	60.48				

Appendix 7:

Conversion of US metal prices from cents per lb to sterling pounds per metric ton* at varying exchange rates

Metal price: cents per lb	Exchange rate: US dollar/£ sterling									
	1.50	1.60	1.70	1.80	1.90	2.00	2.10	2.20	2.30	2.40
					£ sterling per metric ton					
1	14.69	13.77	12.96	12.24	11.60	11.02	10.49	10.01	9.58	9.18
5	73.46	68.85	64.80	61.20	58.00	55.10	52.45	50.05	47.91	45.91
10	146.93	137.70	129.60	122.40	116.00	110.20	104.90	100.10	95.82	91.83
20	293.86	275.40	259.20	244.80	232.00	220.40	209.80	200.20	191.64	183.66
30	440.79	413.10	388.80	367.20	348.00	330.60	314.70	300.30	287.46	275.49
40	587.72	550.80	518.40	489.60	464.00	440.80	419.60	400.40	383.28	367.32
50	734.65	688.50	648.00	612.00	580.00	551.00	524.50	500.50	479.10	459.15
60	881.58	826.20	777.60	734.40	696.00	661.20	629.40	600.60	574.92	550.98
70	1,028.51	963.90	907.20	856.80	812.00	771.40	734.30	700.70	670.74	642.81
80	1,175.44	1,101.60	1,036.80	979.20	928.00	881.60	839.20	800.80	766.56	734.64
90	1,322.37	1,239.30	1,166.40	1,101.60	1,044.00	991.80	944.10	900.90	862.38	826.47
100	1,469.30	1,377.00	1,296.00	1,224.00	1,160.00	1,102.00	1,049.00	1,001.00	958.20	918.30

* metric ton = 2,204 lbs

Appendix 8:

Conversion of tin prices from Malaysian ringgits per pikul* to sterling pounds per metric ton at varying exchange rates

		Exchange rate: Malaysian ringgits/£ sterling			
		3.5	4.0	4.5	5.0
		£ sterling per metric ton			
Malaysian ringgits/pikul	500	2,361	2,066	1,836	1,653
	1,000	4,722	4,132	3,673	3,306
	2,000	9,444	8,264	7,346	6,612
	3,000	14,166	12,396	11,019	9,918

* pikul = 133.33 lbs

Appendix 9:
Useful addresses

Associations

The Aluminium Federation
Broadway House
Calthorpe Road
Birmingham B15 1TN
Tel 021 4550311

Bismuth Institute
Rue Brederode 9
1000 Bruxelles
Belgium

The British Non-Ferrous Metals Federation
6 Bathurst Street
London W1
Tel 01 7237465

Bureau International de la Recuperation
(BIR)
Place du Samedi 13- Bte 4
1000 Bruxelles
Belgium
Tel 2 2178251

CIPEC [Intergovernmental Council of
Copper Exporting Countries]
177 Avenue de Roule
92200 Neuilly sur Seine
Paris
France
Tel 1 7581155

Commodities Research Unit
26 Red Lion Square
London WC1 4RL
Tel 01 2427463

Copper Development Association
Orchard House
Mutton Lane
Potters Bar
Herts
England
Tel 01 7750711

General Services Administration
General Services Building
18th and F St NW
Washington DC 20405
USA
Tel 202 6554000

International Lead/Zinc Study Group
Metro House
58 St James's Street
London SW1
Tel 01 4999373

International Primary Aluminium Institute
9th Floor
New Zealand House
Haymarket
London SW1
Tel 01 9303051

International Tin Council
1 Oxendon Street
London SW1
Tel 01 9300321

Minor Metals Traders Association
69 Cannon Street
London EC4N 5AB
Tel 01 2484444

National Association of Recycling
Industries Inc [NARI]
330 Madison Avenue
New York NY 10017
Tel 212 8677336

Roskill Information Services Ltd
14 College Street
London SW1P 3RZ
Tel 01 2228686

World Bureau of Metal Statistics
6 Bathurst Street
London W2
Tel 01 7237465

Zinc Lead and Cadmium Development
Associations
34 Berkeley Square
London W1
Tel 01 4996636

The Minor Metals Traders' Association

Members

ACLI International
Acli Metal & Ore Division
110 Wall Street
New York
NY 10005, USA
Tel 212 9438700 Telex RCA 232261
D Peterson

Ametalco Trading Limited
29 Gresham Street
London EC2V 7DA
Tel 01 6068800 Telex 885541
J H Jessop

Ayrton & Partners Ltd
Friendly House
21-24 Chiswell Street
London EC1Y 4SN
Tel 01 6385588 Telex 887648
L Lubett

Basmont Metal Co Ltd
Victoria House
Vernon Place
Southampton Row
London WC1
Tel 01 4055065 Telex 27312
I Markson

BOC Minerals
Broadway Chambers
Hammersmith Broadway
London W6
Tel 01 7410661 Telex 934849
B Webb

Brandeis Goldschmidt & Co Ltd
4 Fore Street
London EC2P 2NU
Tel 01 6385877 Telex 84401
I Markson

Brookside Metal BV
Vredenberg 136a
Utrecht
Holland
Tel 30 313851 Telex 40555
G Bromet

Cambridge Metals Limited
7 All Saints Passage
Cambridge CB2
Tel 0223 312111 Telex 817570
F Wolstencroft

Chloride Metals Ltd
Manor Metal Works
Felixstowe Road
Abbey Wood
London SE2
Tel 01 3104444 Telex 896948
J Tilly

Cominco (UK) Ltd
50 Finsbury Square
London EC2A 1BD
Tel 01 6061883 Telex 886563
P Pemberton

Commodity Analysis Ltd
37-39 St Andrews Hill
London EC4V 5DD
Tel 01 2489571 Telex 883356
I McKinnon

Congrain Metals Ltd
Creechurch House
Creechurch Lane
London EC3A 5DA
Tel 01 2830111 Telex 8812396
L Pierce

Continental Metals Corporation
820 Second Avenue
New York
NY 10011
USA
Tel 212 4219811 Telex 425589
D Lehrman

Copalco International Ltd
Suite 354
200 Park Avenue
New York
NY 10017 USA
Tel 212 6977260 Telex 127758
W Veenis

R L Cusick (Metals) Ltd
Cusick House
Church Street
Ware
Herts
Tel 0920 61181 Telex 817395
B Morgan

Entores Ltd
79-83 Chiswell Street
London EC1
Tel 01 6066050 Telex 261932
D S Rogers

Exsud Ltd
Victoria House
Vernon Place
Southampton Row
London WC1
Tel 01 4059862 Telex 264751
J Carson

FLT & Metals Ltd
1-5 Long Lane
London EC1
Tel 01 6061272 Telex 8811917
A Kowalski

Galaxy Metals Corporation
888 Seventh Avenue
New York
NY 10019
USA
Tel 212 5810055 Telex 423462
F Perkiel

Gerald Metals Ltd
Europe House
World Trade Centre
St Katherine-by-the-Tower
London E1 9AA
Tel 01 4810581 Telex 884377
R Kestenbaum

Gill & Duffus Ltd
201 Borough High Street
London SE1 1HW
Tel 01 4077050 Telex 887162, 887588
I M Rankin

Grondmet
Rochussenstraat 125
Rotterdam
Holland
Tel 10 361933 Telex 21689, 25218
E L Dragten

Harlow & Jones Ltd
St Albans House
59 Haymarket
London SW1
Tel 01 8398030 Telex 917691
B Jenkins

Intsel Ltd
83-87 Gracechurch Street
London EC3
Tel 01 6233691 Telex 8811981
R Eschwege

AC Israel Woodhouse Co Ltd
21 Mincing Lane
London EC3R 7DN
Tel 01 6233131 Telex 883136, 883139,
885255
L Hillyard

Lambert Metals Ltd
506-508 Kingsbury Road
London NW9
Tel 01 2049422 Telex 23844
H Masters

Leopold Lazarus Ltd
Gotch House
20-34 St Bride Street
London EC4
Tel 01 5838060 Telex 265544
P Klein

Leigh & Sillavan Ltd
Knights Pools
Windmill Street
Macclesfield
Cheshire SK11 7HR
Tel 0625 31331 Telex 668363
R N W Tolson

Lonconex Ltd
29 Mincing Lane
London EC3
Tel 01 6264383 Telex 885016
N Lowenstein

Maclaine Watson & Co Ltd
2-4 Idol Lane
London EC3R 5DL
Tel 01 2838611 Telex 883854
J H Etherton

Metaleg Metall GmbH
Heinrichstrasse 85
4000 Dusseldorf
West Germany
Tel 211 631011 Telex 8582610
K Krannich

Metallbodio Ltd
Postbox 296
CH-4010 Basel
Switzerland
Tel 61 238953 Telex 62270
P Bodenheimer

Metallgesellschaft AG
British Representative Office
1 Kingsway
London WC2
Tel 01 8364477 Telex 23260
A E T Stroud

Metals for Melting Ltd
Plas Gwyn
Llys Helig Drive
Llandudno
Gwynedd
Wales LL30 2XB
Tel 0492 78935 Telex 61517
P A G Cockrell

Metramet Ltd
Kingswell
58-62 Heath Street
London NW3 1EN
Tel 01 7941131 Telex 25479
J Leviton

Mining & Chemical Products Ltd
Station Wharf
Ealing Road
Alperton
Middlesex
Tel 01 9021191 Telex 923455
N Clayden

Minor Metals Inc
1 Gulf & Western Plaza
Room 918
New York
NY 10023 USA
Tel 212 5418880 Telex 426965
R Weisfisch

Minwood Metals Ltd
Paterson Road
Finedon Industrial Estate
Wellingborough
Northants
Tel 0933 225766 Telex 311394
C Washington

Montanore BV
Vondelstraat 65-67
Amsterdam
Holland
Tel 20 161844 Telex 11063
J Barends

Nemco Metal International Ltd
9 Harrowden Road
Brackmills
Northampton NN4 0BZ
Tel 0604 45404 Telex 826433
R Moule

A J Oster Co
50 Sims Avenue
Providence
Rhode Island USA
Tel 401 4213840 Telex WUD 927747
R M Oster

Powell Metals & Chemicals (UK) Inc
62 Hills Road
Cambridge CB2 1LA
Tel 0223 51775 Telex 817669 Powcang
G Strickland-Clark

Primetal Italia
20124 Milano
Via Cappellini 16
Italy
Tel 2 667124 Telex 33439
Dr M Cassola

Derek Raphael & Co Ltd
116 Gloucester Place
London W1
Tel 01 4872661 Telex 261916
D Raphael

Redlac Metals Ltd
148 Buckingham Palace Road
London SW1
Tel 01 7032276 Telex 888885
J Moore

William Rowland Ltd
Powke Lane
Cradley Heath
Warley
West Midlands B64 5PX
Tel 021 5593031 Telex 337376
D F Brown

Jack Sharkey & Co Ltd
Middlemore Road
Smethwick
Warley
West Midlands
Tel 021 5587444
J Sharkey

Skandinaviska Malm-Och
Metallaktiebolaget
Kungsgatan 6
Box 7547
S-103 93 Stockholm
Sweden
Tel 8 233520 Telex 19552
F Scherlin

Sogemet
161 Avenue Charles de Gaulle
92202 Neuilly sur Seine
France
Tel 1 7471212 Telex 620242
R Lorrillere

Steetley Chemicals
Berk House
PO Box 56
Basing View
Basingstoke
Hants
Tel 0256 29292 Telex 858371
Miss B A C Bailey

Sterling Enterprises Metals Ltd
Sterling House
328 Holloway Road
London N7
Tel 01 6077381 Telex 27325
F Gothold

Tennant Trading Ltd
9 Harp Lane
Lower Thames Street
London EC3
Tel 01 6264533 Telex 884724
A Fergusson

Trans-World Metals Ltd
Walsingham House
35 Seething Lane
London EC3N 4EL
Tel 01 4805701 Telex 8951322
D Reuben

Unimet GmbH
Stahl, Rohre und Metalle
Cecilienallee 21
4000 Dusseldorf 30
Tel 211 450914 Telex 08582622
H-J Jerig

Rene Weil SA
77 Rue de Monceau
75008 Paris
France
Tel 1 5229933-35 Telex 280445
D Amar

Wogen Resources Ltd
17 Devonshire Street
London W1
Tel 01 5805762 Telex 28820
P Robbins

Journals

American Metal Market
7 East 12th Street
New York NY 10003
Tel 212 7414000

Commodity World
Swiss Centre
Leicester Square
London W1
Tel 01 8360775

Metal Bulletin Ltd
45-46 Lower Marsh
London SE1 7RG
Tel 01 6330525

Metals Week
McGraw-Hill Inc
1221 Avenue of the Americas
New York NY 10020
Tel 212 9971221

Tin International
Tin Publications Ltd
7 High Road
London W4 2NE
Tel 01 9959277

London Metal Exchange

The following are member companies of the London Metal Exchange. *These are the ring dealing members.

Amalgamated Metal Corp Ltd
Adelaide House
London Bridge
London EC4R 9DT
Tel 01 6264521 Telex 888701
Cables Amalgameco London

Amalgamated Metal Trading Ltd *
2 Metal Exchange Buildings
Leadenhall Avenue
London EC3V 1LD
Tel 01 6264521 Telex 888704

Ambores Ltd
27 Marylebone Road
London NW1 5JS
Tel 01 4864642

Ametalco Ltd
29 Gresham Street
London EC2V 7DA
Tel 01 6068800 Telex 885541
Cables Nucliform London EC2

Ametalco Merchanting Ltd
1 Noble Street
London EC2V 7DA
Tel 01 6068800 Telex 885541

Ametalco Trading Ltd *
29 Gresham Street
London EC2V 7DA
Tel 01 6068800 Telex 885541
Cables Nucliform London EC2

AM & S Europe Ltd
PO Box 237
1 Redcliff Street
Bristol BS99 7EA
Tel 0272 290421 Telex 449502

Anglo Chemical Metals Ltd *
Gillett House
55 Basinghall Street
London EC2V 5HN
Tel 01 6060811 Telex 886767

Anglo Chemical & Ore Ltd
Gillett House
55 Basinghall Street
London EC2V 5HN
Tel 01 6060811 Telex 885168

Anmercosa Sales Ltd
40 Holborn Viaduct
London EC1P 1AJ
Tel 01 3531545 Telex 264791

Morris Ashby Ltd
10 Philips Lane
London EC3M 8AA
Tel 01 6268071 Telex 886170

Associated Lead Manufacturers Ltd *
Clements House
14-18 Gresham Street
London EC2P 2JS
Tel 01 6064400 Telex 884141
Cables Associated London

Astra Metals & Residues Ltd
Northway House
High Road
Whetstone
London N20 9LP
Tel 01 4464381 Telex 267401

Austral Metals Ltd
9 Park Place
St James's
London SW1A 1LX
Tel 01 4938474 Telex 23672

Ayrton & Partners Ltd
Friendly House
21-24 Chiswell Street
London EC1Y 4SN
Tel 01 6385588 Telex 887648

Bache Halsey Stuart (London) Ltd
Plantation House
Fenchurch Street
London EC3M 3EP
Tel 01 6234646 Telex 883251

Bassett Smith & Co Ltd
9 Harp Lane
Lower Thames Street
London EC3R 6DR
Tel 01 6264533 Telex 884754

Henry Bath & Son Ltd *
Market Buildings
Mincing Lane
London EC3R 7DA
Tel 01 6261981 Telex 887700
Cables Bathota

BICC Ltd
PO Box 1
Prescot
Merseyside L34 5SZ
Tel 0772 4266571 Telex 627364

BICC Metals Ltd
PO Box 1
Prescot
Merseyside L34 5SZ
Tel 0772 4266571 Telex 627364

Billiton-Enthoven Metals Ltd *
Colonial House
Mincing Lane
London EC3R 7DP
Tel 01 6232080 Telex 887939

Billiton Metal & Ores (UK) Ltd
3-4 Lincolns Inn Fields
London WC2A 3AA
Tel 01 8317252 Telex 885737

M C Brackenbury & Co
19 St Mary-at-Hill
London EC3R 8EE
Tel 01 6235701/5707 Telex 883304

Brandeis Goldschmidt & Co Ltd *
4 Fore Street
London EC2P 2NU
Tel 01 6385877 Telex 884401

Brittania Lead Co Ltd
Adelaide House
King William Street
London EC4R 9DX
Tel 01 6267446 Telex 884210

British American Metals Co Ltd
French Railways House
178-179 Piccadilly
London W1V 0QH
Tel 01 4937835 Telex 27380

Cerro Metals (UK) Ltd *
Pinners Hall
Austin Friars
London EC2N 8BE
Tel 01 6285957 Telex 887276
Cables Cerro UK

Chloride Metals Ltd
Felixstowe Road
Abbey Wood
London SE2 9RW
Tel 01 3104444 Telex 896948

Cominco (UK) Ltd*
50 Finsbury Square
London EC2A 1DD
Tel 01 6384000 Telex 886563
Cables Tadanac, London

The Commercial Metal Co Ltd *
Park House
16 Finsbury Circus
London EC2M 7DJ
Tel 01 6385789 Telex 887410, 887419
Cables Cometalco

Commodity Analysis Ltd
37-39 St Andrews Hill
London EC4V 5DD
Tel 01 2489571 Telex 883356

Congrain Metals Ltd
Creechurch House
Creechurch Lane
London EC3A 5DR
Tel 01 2830111 Telex 8812396

Conti Commodity Services Ltd
Europe House
World Trade Centre
London E1 9AA
Tel 01 4883232 Telex 887438
Cables TGR/CCSG

Continental Ore Europe Ltd *
18 Rood Lane
London EC3M 8BS
Tel 01 6233621 Telex 886269

C Czarnikow Ltd
PO Box 602
66 Mark Lane
London EC3P 3EA
Tel 01 4806677 Telex 885011

Charles Davis (Metal Brokers) Ltd *
5 Whittington Avenue
London EC3V 1LE
Tel 01 6263538 Telex 883928
Cables Losange

Delta Metal Co Ltd
1 Kingsway
London WC2B 6XF
Tel 01 8363535 Telex 27762
Cables Delta Kings Ldn

Delta Metal Co Ltd
Greets Green Road
West Bromwich
West Midlands B70 9ER
Tel 021 5536188 Telex 33127
Cables Extrusion West Bromwich

Derby & Co Ltd
Moor House
London Wall
London EC2Y 5JE
Tel 01 6382876 Telex 883801
Cables Plativet London EC2

Dietiker Metals (UK) Ltd
29 Throgmorton Street
London EC2N 2AN
Tel 01 6380911 Telex 887734/35
Cables Dmet UK

Drexel Burnham Lambert Ltd
Winchester House
77 London Wall
London EC2N 1BE
Tel 01 6283200 Telex 884995, 884845
Cables Coburnham London EC2

Enfield Rolling Mills Ltd
Millmarsh Lane
Brimsdown
Enfield
Middlesex
Tel 01 8041255 Telex 22869
Cables Rolen-mil

Engelhard Sales Ltd
St Nicholas House
St Nicholas Road
Sutton
Surrey
Tel 01 6438080 Telex 22317
Cables Platmetals, Sutton, Surrey

H J Enthoven & Sons Ltd
3-4 Lincoln's Inn Fields
London WC2A 3AA
Tel 01 8317252 Telex 885737 BUK G

Entores (Metal Brokers) Ltd *
City Wall House
79-83 Chiswell Street
London EC1Y 4TB
Tel 01 6066050 Telex 887621

EZ Europe Ltd
9 Park Place
St James's
London SW1A 1LX
Tel 01 6291437 Telex 23671

Frank Fehr & Co Ltd
Prince Rupert House
64 Queen Street
London EC4R 1ER
Tel 01 2485066 Telex 888960

Fergusson Wild & Co Ltd
3 St Helen's Place
London EC3A 6DB
Tel 01 5887022 Telex 883503

Gardner Lohmann Ltd
6 Laurence Pountney Hill
London EC4R 0BL
Tel 01 6268086 Telex 884458, 886388

Gerald Metals Ltd *
Europe House
World Trade Centre
St Katherine-by-the-Tower
London E1
Tel 01 4810681 Telex 883869
Cables GMET London

Gill & Duffus Ltd *
St Dunstan's House
^01 Borough High Street
London SE1 1HW
Tel 01 4077050 Telex 887588

Gill & Duffus (Metals) Ltd
St Dunstan's House
201 Borough High Street
London SE1 1HW
Tel 01 4077050 Telex 887588

M Golodetz Ltd
29 Mincing Lane
London EC3R 7EE
Tel 01 6264383 Telex 855016

M Golodetz (Overseas) Ltd
Aldwych House
71-91 Aldwych
London WC2B 4HN
Tel 01 2428888 Telex 23567
Cables Goloseas London WC2

Harlow & Jones Ltd
St Albans House
59 Haymarket
London SW1
Tel 01 8398030 Telex 917691
Cables Harjones

Heathfield Proprietary Ltd
9 Park Place
St James's
London SW1
Tel 01 4938474 Telex 23672

IMI Ltd
PO Box 216
Kynoch Works
Witton
Birmingham B6 7BA
Tel 021 3564848 Telex 366771
Cables Kynoch Birmingham

Imperial Chemical Industries Ltd
Imperial Chemical House
Millbank
London SW1P 3JF
Tel 01 8344444 Telex 21324
Cables Impkemix

Intsel Ltd *
4th Floor
85 Gracechurch Street
London EC3V 0AA
Tel 01 6233691 Telex 8811981
Cables Intsel London EC3

A C Israel Woodhouse Co Ltd
21 Mincing Lane
London EC3P 7DN
Tel 01 6233131 Telex 88742
Cables Acicowood Ldn

Jacob Metals Ltd
8 Waterloo Place
St James's
London SW1Y 4AH
Tel 01 9306953 Telex 918034
Cables Metaljacob

Johnson Mathey Commodities Ltd
5 Lloyds Avenue
London EC3N 3DB
Tel 01 4813181 Telex 884312

Johnson & Nephew (Non-Ferrous) Ltd
Dale House
Tiviot Dale
Stockport SK1 1TP
Tel 061 4775252 Telex 667851

G W Joynson & Co Ltd
14 Trinity Square
London EC3N 4ES
Tel 01 4806921 Telex 885346
Cables Warrior

Leopold Lazarus Ltd *
Gotch House
20-34 St Bride Street
London EC4A 4DL
Tel 01 5838060 Telex 265544
Cables Orminiaz

Leigh & Sillivan Group Ltd
Knights Pool
Windmill Street
Macclesfield
Cheshire SK11 7HR
Tel 0625 31331 Telex 688363
Cables Sillivan Macclesfield

Lewis & Peat (Metals) Ltd
32 St Mary-at-Hill
London EC3R 8DH
Tel 01 6233111 Telex 883983

Lonconex Ltd *
29 Mincing Lane
London EC3R 7EU
Tel 01 6264383 Telex 885016
Cables Lonconex London EC3

J Lowenstein & Co Ltd
Lometa House
12-14 River Road
Barking
Essex IG11 0DG
Tel 01 5911114 Telex 897224
Cables Lometa

Mack Dix & Co Ltd
Plantation House
Mincing Lane
London EC3M 3PH
Tel 01 6260101 Telex 885977

Maclaine Watson & Co Ltd *
2-4 Idol Lane
London EC3R 5DL
Tel 01 2838611 Telex 884328

Metallgesellschaft Ltd *
19-21 Gt Tower Street
London EC3R 5AQ
Tel 01 6264221 Telex 888971
Cables Metallbank London

Metal Scrap and By-Products Ltd
110 Queen Victoria Street
London EC4V 4AU
Tel 01 2368931 Telex 887151

Metdist Ltd *
Barrington House
59-67 Gresham Street
London EC2V 7EY
Tel 01 6060463 Telex 888248

Metro Metal Traders Ltd
Station House
Harrow Road
Wembley HA9 6DE
Tel 01 9034971 Telex 923548

Moutafian Commodities Ltd
2-4 Eastcheap
London EC3M 1AL
Tel 01 6233311 Telex 884412
Cables Moutafian London

Noranda Sales Corp of Canada Ltd
New Zealand House
Haymarket
London SW1Y 4TH
Tel 01 8397601 Telex 24666
Cables Noranda, London

Jeremy Oates Ltd
16 Monument Street
London EC3R 8AJ
Tel 01 6237301 Telex 888264

Pacol Ltd
47 Mark Lane
London EC3R 7QX
Tel 01 6231000 Telex 888712
Cables Pacolanter

Philipp & Lion *
Moor House
London Wall
London EC2Y 5AR
Tel 01 6783060 Telex 888555

Pirelli General Cable Works Ltd
PO Box 4
Western Esplanade
Southampton SO9 7AE
Tel 0703 20381 Telex 47522

J H Rayner (Mincing Lane) Ltd *
50 Mark Lane
London EC3R 7RJ
Tel 01 7099144 Telex 883461

Marc Rich & Co Ltd
49 Wigmore Street
London W1H 9LE
Tel 01 9354455 Telex 21223/4

Richmond Metal Co Ltd
Dellbow Road
North Feltham Trading Estate
Faggs Road
Feltham
Middlesex
Tel 01 8900981 Telex 934857

Henry Rogers (London) Ltd
9 Harp Lane
Great Tower Street
London EC3R 6DP
Tel 01 6264533 Telex 884724

Sharps, Pixley Ltd*
10 Rood Lane
London EC3M 8BB
Tel 01 6238000 Telex 883319

Shearson Hayden Stone Ltd
St Alphage House
2 Fore Street
London EC2Y 5DA
Tel 01 5887505 Telex 888881

Sogemin (Metals) Ltd*
10th Floor
Stock Exchange Building
Throgmorton Street
London EC2N 1EL
Tel 01 6286421 Telex 885244

Tennant Trading (Metals) Ltd*
9 Harp Lane
Lower Thames Street
London EC3R 6DR
Tel 01 6264533 Telex 887604

H P Thompson & Sons Ltd*
8 Lloyds Avenue
London EC3N 3AB
Tel 01 4807951 Telex 8951635
Cables Knilomet London EC3

Triland Metals Ltd*
Bow Bells House
Bread Street
London EC4M 9BQ
Tel 01 2365551 Telex 888677
Cables Trimet, London EC4

Wallace Brothers Commodities Ltd
108 Fenchurch Street
London EC3M 5HP
Tel 01 4818671 Telex 887442

H A Watson & Co Ltd
11 Mansfield Road
Rotherham
Yorkshire
Tel 0709 74293 Telex 547393

Westralian Ores and Metals Ltd
15 Blackfriars Lane
London EC4V 6ER
Tel 01 2485987 Telex 887444

Wilson Smithett and Cope*
32 St Mary-at-Hill
London EC3R 8DH
Tel 01 6239333 Telex 883898
Cables Infados

Wogen Resources Ltd
17 Devonshire Street
London W1N 1FS
Tel 01 5805762 Telex 28820

Woodhouse Drake and Carey Ltd
Three Quays
Tower Hill
London EC3R 6EP
Tel 01 6237575 Telex 886271
Cables Progne

Rudolf Wolff and Co Ltd*
Knollys House
11 Byward Street
London EC3R 5ED
Tel 01 6268765 Telex 885034
Cables Occasion, London

Commodity Exchange Inc (COMEX), New York

The following are trading members of
Comex as at March 1978

M Abrams
120 East Hartsdale Avenue
Hartsdale, New York, NY 10530

I Afram
Afram Bros Inc
900 South Water Street
Milwaukee, Wisconsin 53204

J Arkin
300 South Wacker Drive
Chicago, Illinois 60606

R Arias
Mintz, Marcus & Co
4 World Trade Center
8th Floor
New York, NY 10048

R Aron
J Aron & Co Inc
160 Water Street
New York, NY 10038

M Ashby
Minemet Metals Inc
450 Park Avenue
New York, NY 10022

M Ashraf
c/o Eastern Trade Corp
Sheikha Building
PO Box 1376
Deira, Dubai
United Arab Emirates

D F Auciello
Great American Trading Co
4 World Trade Center
8th Floor
New York, NY 10048

V W Augello
674 Keil Street
Elmont, NY 11003

J J Baer
Northern Investment SA
PO Box 14-330
Madrid, Spain

H F Baerwald
International Minerals & Metals Corp
919 Third Avenue
New York, NY 10022

M R Baldini
Baldini, Proffitt, Reilly & Co
61 Broadway, Room 2526
New York, NY 10006

R Barba
Clayton Brokerage Co of St Louis
No 2 Gate Field Drive
Greenwich, Connecticut 06830

F W Baur
Harmon, Lichtenstein & Co
4 World Trade Center
Room 7170
New York, NY 10048

H Becker-Fluegel
NRT Metals Inc
576 Fifth Avenue
New York, NY 10036

S Bell
Stanley B Bell & Co
4 World Trade Center, 8th Floor
New York, NY 10048

M Belmont
4 World Trade Center, 8th Floor
New York, NY 10048

K G Bergman
Amax Copper Inc
200 Park Avenue
New York, NY 10017

V Besso
Intsel Corp
825 3rd Avenue
New York, NY 10022

L Bilello
Stanley B Bell & Co
4 World Trade Center, 8th Floor
New York, NY 10048

G H Bischoff
Sharps, Pixley Inc
100 Wall Street
New York, NY 10005

L J Blankenship
Shearson Hayden Stone Inc
4 World Trade Center, 8th Floor
New York, NY 10048

P B Bleecker
Scholtz & Co Inc
110 Wall Street
New York, NY 10005

R Bleichroeder
Samuel Montagu & Co Ltd
114 Old Broad Street
London EC2 2HY, England

E J Blot
1075 North Ocean Blvd
Palm Beach, Florida 33480

L J Blum
Eisen & Blum Inc
141 West Jackson Blvd
Chicago, Illinois 60604

G Boglione
Boys, Investment & Trade Ltd
AV 15, App 112
Auzere, Sion, Switzerland

O Boonshoft
1330 Winters Bank Tower
Dayton, Ohio 45402

J P Branco
Floor Broker Associates
4 World Trade Center, 8th Floor
New York, NY 10048

J J Brandler
Lee B Stern & Co Ltd
4 World Trade Center, 8th Floor
New York, NY 10048

R J Bresky
Thomson McKinnon Securities Inc
One New York Plaza
New York, NY 10004

S J Briggs
Floor Broker Associates
4 World Trade Center, 8th Floor
New York, NY 10048

A Brodsky
4 World Trade Center, 8th Floor
New York, NY 10048

E J Browdy
c/o Anglo American Warehouse Inc
40th Street & Packers Avenue
Chicago, Illinois 60609

V Buccellato
B & B Commodities
4 World Trade Center
New York, NY 10048

J C Buck
Asarco Inc
120 Broadway
New York, NY 10005

O Burchard
The Ore & Chemical Corp
605 Third Avenue
New York, NY 10016

M Burke
Smith Barney, Harris Upham & Co Inc
1345 Avenue of the Americas
New York, NY 10020

M A Buscemi
4 World Trade Center, 8th Floor
New York, NY 10048

D R Campbell
Amalgamet Inc
733 Third Avenue
New York, NY 10017

A Capinegro
Lincolnwood Inc
141 West Jackson Blvd
Chicago, Illinois 60604

C R Caruana
Dean Witter Reynolds Organization Inc
Two Broadway
New York, NY 10004

E C Cashman
Cashman & Co
141 West Jackson Blvd
Chicago, Illinois 60604

S Cherwin
2803 Covered Bridge Road
Merrick, NY 11566

S Chilewich
New York Hide Trading Co
120 Wall Street
New York, NY 10005

P A Cohen
Herzfeld & Stern
30 Broad Street
New York, NY 10004

J R Collins
Greene & Collins Inc
175 West Jackson Blvd
Chicago, Illinois 60604

D R Comenzo
c/o New York Cotton Exchange
4 World Trade Center, 8th Floor
New York, NY 10048

J J Conheeney
Merrill Lynch, Pierce, Fenner & Smith Inc
One Liberty Plaza
165 Broadway
New York, NY 10006

F J Coppola
Merrill Lynch, Pierce, Fenner & Smith Inc
One Liberty Plaza
165 Broadway
New York, NY 10006

H J Coyne
J Aron & Co Inc
160 Water Street, 19th Floor
New York, NY 10038

M L Coyne
J Aron Commodities Corp
160 Water Street, 19th Floor
New York, NY 10038

N S Crystal
PO Box 738
Union City, New Jersey 07087

P H Dalton Jr
Interstate Securities Corp
221 South Tryon Street
Charlotte, North Carolina 28202

J R D'Anna
Shearson Hayden Stone Inc
4 World Trade Center
New York, NY 10048

R P D'Arcy
Amax Copper Inc
200 Park Avenue
New York, NY 10017

B S Davis III
Davis & Co
1180 NCNB Plaza
Charlotte, North Carolina 28280

C D E Davis
Lonconex Ltd
29 Mincing Lane
London EC3 RFEU

N L Deak
Perera Company Inc
29 Broadway
New York, NY 10006

C H Deane
Deane Commodity Corp
4 World Trade Center
Box 289
New York, NY 10048

M Decker
Bear, Stearns & Co
55 Water Street
New York, NY 10041

A Del Re
4 World Trade Center, 8th Floor
New York, NY 10048

R DeMartino
Commodity International Co
84 William Street
New York, NY 10038

S DeMicoli
Cerro Sales Corp
250 Park Avenue
New York, NY 10017

M H Denberg
4 World Trade Center, 8th Floor
New York, NY 10048

R F DeVellis
Federbush & Sabin
4 World Trade Center, 8th Floor
New York, NY 10048

T L Diamond
T L Diamond & Company Inc
30 Rockefeller Plaza
New York, NY 10020

R Digiacomo
Loeb Rhoades Hornblower & Co
42 Wall Street
New York, NY 10015

C Diliberto
E Modet & Co Inc
4 World Trade Center, 8th Floor
New York, NY 10048

J Diliberto
Macro International Group Inc
4 World Trade Center, 8th Floor
New York, NY 10048

T H Dittmer
Ray E Friedman & Co
222 South Riverside Plaza
Chicago, Illinois 60606

R D Donchian
Shearson Hayden Stone Inc
Two Greenwich Plaza
Greenwich, Connecticut 06830

J P Donnelly Jr
Balfour, MacLaine Inc
88 Pine Street
New York, NY 10005

G E Doty
Goldman, Sachs & Co
55 Broad Street
New York, NY 10004

P G Drogoul
Minemet Metals Inc
450 Park Avenue
New York, NY 10022

S S Duberstein
Redel Trading Co Inc
140 Cedar Street
New York, NY 10006

F A Dubie
4 World Trade Center, 8th Floor
New York, NY 10048

Lord Aldenham A Durant
R J Rouse & Co Ltd
23 Blomfield Street
London EC2M 7NL,

D R Dustin
Precious Metals Corp of America
5 World Trade Center, Suite 7229
New York, NY 10048

C Edelstein
J Aron Commodities Corp
4 World Trade Center
New York, NY 10048

G A Eichenberg
General Hide & Skin Corp
11 Park Place
New York, NY 10007

H G Eisenberg
Brandeis, Goldschmidt & Co Inc
919 Third Avenue
New York, NY 10022

I Ergas
1199 Park Avenue
New York, NY 10028

J J Ezzo
Mintz, Marcus & Co
4 World Trade Center, 8th Floor
New York, NY 10048

P K Falkenrath
Intsel Commodities Inc
825 Third Avenue
New York, NY 10022

S Fashena
Loeb Rhoades Hornblower & Co
42 Wall Street
New York, NY 10005

C Federbush
Federbush & Sabin
225 East 73rd Street
New York, NY 10021

M A Fellman
Dean Witter Reynolds Organization Inc
Two Broadway
New York, NY 10004

H D Ferer
Aaron D Ferer & Sons Co
909 Abbot Drive
Omaha, Nebraska 68102

D A Ferretti
Insilco Corp
500 South Broad Street
Meriden, Connecticut 06450

S H Fields
First Wall Street Settlement Corp
40 Wall Street
New York, NY 10005

R J Fisher
4 World Trade Center, 8th Floor
New York, NY 10048

M P Fitzgerald
Balfour, MacLaine Inc
785 Poole Avenue
Hazlet, New Jersey 07730

T P Fletcher
161-10 32nd Avenue
Flushing, NY 11358

F J Forlenza Sr
Felix J Forlenza & Co
16 West Granada Drive
Bucktown, New Jersey 08723

F J Forlenza Jr
Felix J Forlenza & Co
4 World Trade Center, 8th Floor
New York, NY 10048

L Forlenza
4 World Trade Center, 8th Floor
New York, NY 10048

S Forlenza
101 N Carmelina Avenue
Los Angeles, California 90049

E L Frank
Engelhard Minerals & Chemical Corp
Philipp Bros Div
1221 Avenue of the Americas
New York, NY 10020

A L Freeman
Rosenthal & Co
4 World Trade Center
New York, NY 10048

A Fried Jr
Albert Fried & Co
77 Water Street
New York, NY 10005

K A Friedlander
Brandeis, Goldschmidt & Co Inc
919 Third Avenue
New York, NY 10022

H M Frietsch
International Systems & Controls Corp
2727 Allen Parkway
Houston, Texas 77019

M A Fromer
Lewis & Peat Trading Inc
460 Park Avenue
New York, NY 10022

G J Gange
UV Industries Inc
437 Madison Avenue
New York, NY 10022

G P Garbe
588 Rose Boulevard
Baldwin, NY 11510

R L Gardner
4 World Trade Center, 8th Floor
New York, NY 10048

G Garfield
Garfield & Co
306 Carter Road
Princeton, New Jersey 08540

D J Garrison
90 Kingfisher Drive
Middletown, New Jersey 07748

D Garrity
General Cable Corp
500 West Putnum Avenue
Greenwich, Connecticut 06830

A G Gero
Drexel Burnham Lambert Inc
4 World Trade Center, 8th Floor
New York, NY 10048

J Gill
Case Postale 159
1012 Lausanne
Chailly, Switzerland

M L Ginzberg
Primary Industries Corp
666 Fifth Avenue
New York, NY 10019

M Glassberg
Ag-Met Inc
PO Box 523
Hazleton, Pennsylvania 18201

B I Goldberg
J Aron Commodities Corp
4 World Trade Center
New York, NY 10048

H Goodman
7851 South Ashland Avenue
Chicago, Illinois 60620

M J Goodman
54 Levering Circle
Bala Cynwyd, Pennsylvania 19004

W D Goodman
Freehling & Co
120 South La Salle Street
Chicago, Illinois 60603

N I Gosberg
N I Gosberg & Co Inc
One Horizon Road, Apt 1020
Fort Lee, New Jersey 07024

A P Graham
NMR Metals Inc
One Rockefeller Plaza
New York, NY 10020

J S Granger
Granger & Company
111 Broadway
New York, NY 10006

L Gray
Gray International Inc
1271 Avenue of the Americas
New York, NY 10020

H Green
44 East 67th Street
New York, NY 10021

M Greenberg
4 World Trade Center, 8th Floor
New York, NY 10048

S Greenberg
Bache Halsey Stuart Shields Inc
100 Gold Street
New York, NY 10038

R G Gremald
Ametalco Inc
530 Fifth Avenue
New York, NY 10036

R S Griffin
141 West Jackson Blvd
Room 1580
Chicago, Illinois 60604

H Grillo
Wilhelm Grillo Handelsgesellschaft
AM Grillo Park 5
Duisborg Hamborn, West Germany

M A Grisez
Billiton Metals & Ores USA Inc
600 Fifth Avenue
New York, NY 10020

E R Haboush
Galaxy Diamond Corp
48 West 48th Street
New York, NY 10036

P H Halladin
85 Hillcrest Drive
Upper Saddle River, New Jersey 07458

H A Handleman
4 World Trade Center, 8th Floor
New York, NY 10048

J H Hanemann
Felix J Forlenza & Co
4 World Trade Center, 8th Floor
New York, NY 10048

R B Harmon
Harmon, Lichtenstein & Co
4 World Trade Center, Room 7170
New York, NY 10048

D Harris
Richardson Securities Inc
40 Wall Street
New York, NY 10005

H U Harris Jr
Smith Barney, Harris Upham & Co Inc
1345 Avenue of the Americas
New York, NY 10020

J E Harris
150 East Palmetto Park Road
Boca Raton, Florida 33432

C Hatten
ACLI International Commodity Services Inc
110 Wall Street
New York, NY 10005

S Haydak
Easton & Co
4 World Trade Center, 8th Floor
New York, NY 10048

H J Heinold
Heinold Commodities Inc
222 South Riverside Plaza
Chicago, Illinois 60606

R Helfer
Helfer Commodities Corp
5 World Trade Center
New York, NY 10048

R Hendricks
Hendricks & Co
4 World Trade Center, Box 324
New York, NY 10048

J J Herman
c/o New York Cotton Exchange
4 World Trade Center, 8th Floor
New York, NY 10048

R J Herstein
4 World Trade Center, 8th Floor
New York, NY 10048

A Hettena
Republic Clearing Corp
450 Fifth Avenue
New York, NY 10018

L A Hockstader II
Rothschild, Unterberg, Towbin, LF
55 Water Street
New York, NY 10041

E W Hoffstatter Jr
Sharps, Pixley Inc
100 Wall Street
New York, NY 10005

J J Hohenberg
The Hohenberg Company
PO Box 193
Memphis, Tennessee 38013

F F Horn
Dean Witter Reynolds Organization Inc
Two Broadway
New York, NY 10004

J A Hunter
Peavey Company
2550 Board of Trade Building
Chicago, Illinois 60604

A C Israel
ACLI International Inc
110 Wall Street
New York, NY 10005

R Jacobson
Reading Industries Inc
789 Downing Street
Teaneck, New Jersey 07666

Dr H G Jarecki
Mocatta Metals Corp
25 Broad Street
New York, NY 10004

Dr R W Jarecki
Brody, White & Co Inc
25 Broad Street
New York, NY 10004

D M Jenkins
London, Polanish Jenkins Inc
4 World Trade Center, 8th Floor
New York, NY 10048

D T Johnston
E F Hutton & Co Inc
One Battery Park Plaza
New York, NY 10004

J W Johnston
Gold Fields American Corp
230 Park Avenue
New York, NY 10017

D E Jones
4 World Trade Center, 8th Floor
New York, NY 10048

D Jones
Arrow Manufacturing Co
885 Freeway Drive North
Columbus, Ohio 43229

P P Jones
Floor Broker Associates
4 World Trade Center, 8th Floor
New York, NY 10048

E J Karmin
Mintz, Marcus & Co
4 World Trade Center, 8th Floor
New York, NY 10048

E Katz
Adolph Hirsch & Co
500 Fifth Avenue, Suite 1724
New York, NY 10036

G Kaufman
4 World Trade Center, 8th Floor
New York, NY 10048

B Kessner
67 Navajo Avenue
Lake Hiawatha, New Jersey 07034

H S Kipnis
H S Kipnis & Co
209 South LaSalle Street
Chicago, Illinois 60604

M Klein
Klein & Co
170 Broadway, Room 806
New York, NY 10006

R Klein
Klein & Co
170 Broadway, Room 806
New York, NY 10006

A M Kohl
Woodland Investment Partners
141 West Jackson Blvd, Room 2002
Chicago, Illinois 60604

A Kraus
Continental Copper & Steel Industries Inc
12 Commerce Drive
Cranford, New Jersey 07016

I Kreitner
Wark Trading Co
4 World Trade Center
New York, NY 10048

I W Kuhlik
4 World Trade Center, 8th Floor
New York, NY 10048

M D Kurzrok
Lehman Brothers Kuhn Loeb Inc
55 Water Street
New York, NY 10041

R G LaFroscia
Ferriso, LaFroscia & Co
4 World Trade Center, 8th Floor
New York, NY 10048

W T Lakin
4 World Trade Center, 8th Floor
New York, NY 10048

D M Lambert
E P Lambert Co
1416 First National Tower
Akron, Ohio 44308

A H Lamborn III
Shearson Hayden Stone Inc
4 World Trade Center, 8th Floor
New York, NY 10048

W L Larson
Cargill Investor Services Inc
2 Broadway
New York, NY 10004

W C Layton Jr
Mine Brook Road
Far Hills, New Jersey 07931

M Lazarus
4 World Trade Center, 8th Floor
New York, NY 10048

L H Leathers
Shearson Hayden Stone Inc
One Western Union International Plaza
New York, NY 10004

G J Leinsdorf
Philipp Brothers Div of
Engelhard Minerals & Chemicals Corp
1221 Avenue of the Americas
New York, NY 10020

N S Leist
c/o New York Mercantile Exchange
4 World Trade Center
New York, NY 10048

G L Lennard
Gerald Metals Inc
One Battery Park Plaza
New York, NY 10004

R Levenson
Intramet Corporation
437 Madison Avenue
New York, NY 10022

M H Levine
Mark H Levine & Co
4 World Trade Center, 8th Floor
New York, NY 10048

L Levy
Oppenheimer & Co Inc
One New York Plaza
New York, NY 10004

G M Levyns
Continental Grain Co
277 Park Avenue
New York, NY 10017

E Leytess
Leytess Metal & Chemical Corp
500 Fifth Avenue
New York, NY 10036

G Lichtenstein
4 World Trade Center, 8th Floor
New York, NY 10048

C Lichtman
J Lichtman & Sons
241 Freylinghuysen Avenue
Newark, New Jersey 07114

M Liebman
M Lib/Ras International
222 South Riverside Plaza
Chicago, Illinois 60606

N Lion
Phillip & Lion
Moor House, London Wall
London EC2Y 5AR, England

A Lipper III
Arther Lipper Corp
176 Broadway
New York, NY 10038

F A Lisi
4 World Trade Center, 8th Floor
New York, NY 10048

N J Lisi
4 World Trade Center, 8th Floor
New York, NY 10048

F A Lissauer
Associated Metals & Minerals Corp
733 Third Avenue
New York, NY 10017

M W Lissner
Lissner Minerals & Metals Inc
1000 N North Branch Street
Chicago, Illinois 60622

J L Loeb Jr
Loeb Rhoades Hornblower & Co
375 Park Avenue
New York, NY 10022

R E Loewenberg
R E Loewenberg Capital Management Corp
450 Park Avenue
New York, NY 10022

A O London
4 World Trade Center, 8th Floor
New York, NY 10048

A Lotterman
Delos Commodities Inc
4 World Trade Center
New York, NY 10048

A G Lowenthal
Cowen & Co
One Battery Park Plaza
New York, NY 10004

R E McDonnell
Commodity International Co
84 William Street
New York, NY 10038

J M McHale
4 World Trade Center, 8th Floor
New York, NY 10048

M L Maduff
Maduff & Sons
222 Riverside Plaza
Chicago, Illinois 60606

W Malinchak
4 World Trade Center, 8th Floor
New York, NY 10048

C J Maneri
70 Fox Hunt Crescent
Oyster Bay Cove
PO Syosset, NY 11791

J Manne
839 Ashland Avenue
River Forest, Illinois 60305

H R Marcus
Mintz, Marcus & Co
4 World Trade Center, 8th Floor
New York, NY 10048

E Margulies
Lonray (Sugar) Inc
77 Water Street
New York, NY 10005

S C Markoff
A-Mark Trading Corp
United California Bank Building
9601 Wilshire Blvd, Suite 25
Beverly Hills, California 90210

N Marone
E Modet & Co Inc
24 Lundi Court
Staten Island, NY 10314

M Marx
United Equities Commodities Co
160 Broadway
New York, NY 10038

G A Mattana
27 Partridge Drive
Roslyn, NY 11576

C Mattey
Bache Halsey Stuart Shields Inc
100 Gold Street
New York, NY 10038

L Mayer
Mayer & Schweitzer Inc
30 Montgomery Street
Jersey City, New Jersey 07302

M Mehl
4 World Trade Center, 8th Floor
New York, NY 10048

D W Merriman
151 East 80th Street
New York, NY 10021

D W Mesker
A G Edwards & Sons Inc
One North Jefferson
St Louis, Missouri 63103

C Miller
Paine, Webber, Jackson & Curtis Inc
25 Broad Street
New York, NY 10004

H Mintz
2200 Park Lane
Hollywood, Florida 33021

L A Mintz
Mintz, Marcus & Co
4 World Trade Center, 8th Floor
New York, NY 10048

J E Moore
Harmon, Lichtenstein & Co
4 World Trade Center, Room 7170
New York, NY 10048

J F Moore
Harmon, Lichtenstein & Co
4 World Trade Center, Room 7170
New York, NY 10048

J E P Moore
J Aron & Co
160 Water Street
New York, NY 10038

J L Morace
457 Ridge Road
Mahwah, New Jersey 07430

F L Morosco
TransWorld Metals
770 Lexington Avenue
New York, NY 10021

O J Morrissey
500 East 77th Street, Apt 2025
New York, NY 10021

R A Moskowitz
Ram Commodities Inc
4 World Trade Center
New York, NY 10048

K F Mountcastle Jr
Dean Witter Reynolds Organization Inc
120 Broadway
New York, NY 10005

S B Mudd
Handy & Harman
850 Third Avenue
New York, NY 10022

E Mugdan
OCC Commodity Inc
605 Third Avenue
New York, NY 10016

J M Murray
E F Hutton & Co Inc
4 World Trade Center, 8th Floor
New York, NY 10048

E Nadler
126 East 61st Street
New York, NY 10021

A Napoletano
4 World Trade Center
New York, NY 10048

N L Nathanson
N L Nathanson & Co
460 North Street
Greenwich, Connecticut 06830

L B Nelson
C Tennant Sons & Co of NY
100 Park Avenue
New York, NY 10017

M H Nelson
4 World Trade Center, 8th Floor
New York, NY 10048

R Nessim
Philip Bros Div of
Engelhard Minerals & Chemical Corp
1221 Avenue of the Americas
New York, NY 10020

H Neu
Hugo Neu & Sons Inc
380 Madison Avenue, Suite 1710
New York, NY 10017

E G Newman
Chilewich Sons & Co
120 Wall Street
New York, NY 10005

J Nordlicht
4 World Trade Center, 8th Floor
New York, NY 10048

J A Novak
16-66 Bell Blvd
Bayside, NY 11360

T P O'Hare
Merrill Lynch, Pierce, Fenner & Smith Inc
One Liberty Plaza
165 Broadway
New York, NY 10006

E J Oliveri
4 World Trade Center, 8th Floor
New York, NY 10048

C Oliverio
4 World Trade Center, 8th Floor
New York, NY 10048

J M Ondo
4 World Trade Center, 8th Floor
New York, NY 10048

J Orefice
c/o Commodity Exchange Inc
4 World Trade Center
New York, NY 10048

M Parsoff
Lee B Stern & Co Ltd
141 West Jackson Blvd
Chicago, Illinois 60604

D E Peiser
Kawama Securities Corp
50 Broadway, Room 1801
New York, NY 10004

R N Peters
Peters & Co
141 West Jackson Blvd
Chicago, Illinois 60604

M Piazza
Mintz, Marcus & Co
4 World Trade Center, 8th Floor
New York, NY 10048

J J Place
Merrill Lynch Pierce Fenner & Smith Inc
165 Broadway
One Liberty Plaza
New York, NY 10006

R C Podewils
C Tennant Sons & Co of NY
100 Park Avenue
New York, NY 10017

F Polanish
London, Polanish Jenkins Inc
4 World Trade Center
New York, NY 10048

J R Poulat
Sociedad Financiera De Industria y
Descuento SA
Av Fco I, Madero No 42, PB
Mexico, DF, Mexico

H A Pratt
Cerro Sales Corp
250 Park Avenue
New York, NY 10017

B Pressner
Pressner Trading Corp
4 World Trade Center, 8th Floor
New York, NY 10048

A G W Prockter
Billiton-Enthoven Metals Ltd
Colonial House, Mincing Lane
London EC3R 7DR, England

B Quint
Easton & Co
4 World Trade Center, 8th Floor
New York, NY 10048

L Rabinowitz
Bear, Stearns & Co
55 Water Street
New York, NY 10041

A Racz
Philips, Appel & Walden Inc
111 Broadway
New York, NY 10006

C E Rasher
McMahon Iron & Metal Co Inc
3300 Conner Street
Bronx, NY 10475

I Redel
Redel Trading Co Inc
140 Cedar Street
New York, NY 10006

N A Redel
Redel Trading Co Inc
140 Cedar Street
New York, NY 10006

O F Rehders
c/o New York Cotton Exchange
4 World Trade Center, 8th Floor
New York, NY 10048

E Reynolds
Spiral Metal Company Inc
515 Madison Avenue
New York, NY 10022

M Rich
Marc Rich & Co Inc
280 Park Avenue
New York, NY 10017

A J Ritholtz
c/o New York Cotton Exchange
4 World Trade Center, 8th Floor
New York, NY 10048

A A Rosa
Rosa Commodities
4 World Trade Center
New York, NY 10048

S Rosenberg
425 East 58th Street
New York, NY 10022

F Rosenblatt
Primary Metal & Mineral Corp
666 Fifth Avenue
New York, NY 10019

M B Rosenthal
9 Steven Lane
Kings Point, NY 11024

S Rothberg
N 1406 Park Towne Place
Philadelphia, Pennsylvania

R E Rowland
4 World Trade Center, 8th Floor
New York, NY 10048

D S Rubin
Thomson McKinnon Securities Inc
One New York Plaza
New York, NY 10004

A J Rudner
Felix J Forlenza & Co

A J Rudner
Felix J Forlenza & Co
4 World Trade Center, 8th Floor
New York, NY 10048

H Sacks
Brandeis, Goldschmidt & Co (Commodities)
Inc
919 Third Avenue
New York, NY 10022

R Saitta
Bache Halsey Stuart Shields Inc
4 World Trade Center
New York, NY 10048

J A Samuel
J A Samuel & Co
233 Broadway
New York, NY 10007

J N Santagata
Mintz, Marcus & Co
4 World Trade Center, 8th Floor
New York, NY 10048

E Santamaria
E Santamaria & Co Inc
PO Box 2015
Manila, PI

P Sarpi
c/o New York Cocoa Exchange Inc
30 Fulton Street
New York, NY 10038

A H Schmidt-Fellner
Amax Inc
2 Greenwich Plaza
Greenwich, Connecticut 06830

A Schneiderman
180 Lakeside Drive So
Lawrence, NY 11559

M Schnell
Mel Schnell Co
4 World Trade Center, 8th Floor
New York, NY 10048

F G Schreiber
Lambros Metals Inc
2 West 45th Street
New York, NY 10036

M L Schultz
ACLI Metal & Ore Co
110 Wall Street
New York, NY

H Schwartz
140 Poplar Drive
Roslyn, NY 11576

M Sepenuk
21 Hyatt Avenue
Newark, New Jersey 07105

N N Shak
22 Riverside Drive Apt A-2
Cranford, New Jersey 07016

I E Shein
ACLI International Inc
110 Wall Street
New York, NY 10005

R Sicuranza
4 World Trade Center, 8th Floor
New York, NY 10048

T J Silk
4 World Trade Center, 8th Floor
New York, NY 10048

A P Silverberg
4 World Trade Center, 8th Floor
New York, NY 10048

B Sirota
Genora Trading Company Inc
300 71st Street, Room 440
Miami Beach, Florida 33141

N L Sirota
115 Broadway, Room 1818
New York, NY 10006

E G Smith
Albert Trostel & Sons Co
800 N Marshall Street
Milwaukee, Wisconsin 53202

J J Smith
Fahnestock & Co
110 Wall Street
New York, NY 10005

A S Snyder
TransWorld Metals Inc
770 Lexington Avenue
New York, NY 10022

I Solomon
ACLI International Commodity Services
Corp
110 Wall Street
New York, NY 10005

J Solta
4455 Douglas Avenue
Riverdale, NY 10463

J M Soriano
Ansor Corporation
1351 Washington Blvd
Stamford, Connecticut 06902

A C Speyer Jr
1202 Benedum Trees Building
Pittsburgh, Pennsylvania 15222

J S Spiller
Mintz, Marcus & Co
4 World Trade Center, 8th Floor
New York, NY 10048

L W Steib
1647 Burnett Street
Brooklyn, NY 11229

E A Stein
4 World Trade Center, 8th Floor
New York, NY 10048

R R Stemland
Paine, Webber, Jackson & Curtis Inc
25 Broad Street, 19th Floor
New York, NY 10004

A Steo
Shearson Hayden Stone Inc
4 World Trade Center, 8th Floor
New York, NY 10004

J Stern
Stern & Stern Securities Inc
110 Wall Street
New York, NY 10005

L B Stern
Lee B Stern & Co Ltd
141 West Jackson Blvd
Chicago, Illinois 60604

J F Stewart
Loeb Rhoades Hornblower & Co
72 West Adams Street
Chicago, Illinois 60603

D A Suskind
J Aron & Co Inc
160 Water Street, 19th Floor
New York, NY 10038

S Tanne
251 Central Park West
New York, NY 10024

C P Tant
Truebner & Co Inc
130 John Street
New York, NY 10038

R A H Tejani
TransWorld Agencies
PO Box 648
Dubai, United Arab Emirates

C Terrein
Financiera Banamex SA
Isabel La Catolica No 39
Mexico 1, DF

E F Thaute
E F Hutton & Co Inc
One Battery Park Plaza
New York, NY 10004

T Thomte
Thomte & Co Inc
47 Commercial Wharf
Boston, Massachusetts 02110

P N Tuomey
4 World Trade Center, 8th Floor
New York, NY 10048

D C Turman
4 World Trade Center, 8th Floor
New York, NY 10048

P D F Varrall
Johnson Matthey Bankers Ltd
5 Lloyds Avenue
London EC3N 3DB, England

J Voe
Rosenthal Equities Inc
5 World Trade Center, Suite 6411
New York, NY 10048

J Vuillequez
Metal Traders Inc
60 East End Avenue
New York, NY 10028

J A Wade Jr
One Beekman Place
New York, NY 10022

J Wagner
1957 Kimball Street
Brooklyn, NY 11234

M R Wagner
1957 Kimball Street
Brooklyn, NY 11234

H A Wall
Harvey Wall Company
600 Avenue Conde 'F'
Coral Gables, Florida 33134

L S Walsh
4 World Trade Center, 8th Floor
New York, NY 10048

R Walton
4 World Trade Center, 8th Floor
New York, NY 10048

N D Waltuch
Conti Commodity Services Inc
4 World Trade Center, Suite 774
SE Plaza Building
New York, NY 10048

A Weeks
E Modet & Co
4 World Trade Center, 8th Floor
New York, NY 10048

L H Weiss
15 Percheron Lane
Roslyn Heights, NY 11577

H J Weissblatt
Brody, White & Co Inc
25 Broad Street
New York, NY 10004

P Weissman
Liberty Smelting Works (1962) Ltd
PO Box 840, Postal Station A
Montreal Quebec, Canada H3C 2V5

H C Weltzein Jr
209 Bear Ridge Road
Pleasantville, NY 10570

G J Westheimer
4 World Trade Center, 8th Floor
New York, NY 10048

S Westheimer
6080 Northwest 44th Street
Lauderhill, Florida 33313

P I Willensky
Loeb Rhoades Hornblower & Co
42 Wall Street
New York, NY 10005

R Willstatter
Gerald Commodity Services Inc
One Battery Park Plaza
New York, NY 10004

R Winzap
Bullion Exchange & Trading Co Ltd
PO Box 2353, CH 1002 Lausanne
4 Avenue De La Gare
CH 1003 Lausanne, Switzerland

I Witkin
82 Wall Street
New York, NY 10005

W I Witkin
113 Campfire Road
Chappaqua, NY 10514

F F

F F Wolff
Rudolf Wolff & Co Ltd
Knollys House
11 Byward Street
London EC3, England

G E Woodburn
Precious Metals Corp of America
5 World Trade Center, Suite 7229
New York, NY 10048

S Yalkowsky
891 Sheridan Avenue
Bronx, NY 10451

E J Young Jr
Clayton Brokerage Co of St Louis Inc
7701 Forsyth Blvd
Clayton, Missouri 63105

B C Zins
4 World Trade Center, 8th Floor
New York, NY 10048

Markets dealing in non-ferrous metals

Commodity Exchange Inc
4 World Trade Centre
New York NY 10048
Tel 212 9382222
Cables Comexinc

Datuk Keramat Smelting Sdn Berhad
as joint operators of Straits Tin Market
73 Jalan Datuk Keramat
PO Box No 280
Penang
Malaysia

The London Metal Exchange
Whittington Ave
London EC3V 1LB, UK
Tel 01 6261011
Cables Metma London EC3